SpringerBriefs in Geography

SpringerBriefs in Geography presents concise summaries of cutting-edge research and practical applications across the fields of physical, environmental and human geography. It publishes compact refereed monographs under the editorial supervision of an international advisory board with the aim to publish 8 to 12 weeks after acceptance. Volumes are compact, 50 to 125 pages, with a clear focus. The series covers a range of content from professional to academic such as: timely reports of state-of-the art analytical techniques, bridges between new research results, snapshots of hot and/or emerging topics, elaborated thesis, literature reviews, and in-depth case studies.

The scope of the series spans the entire field of geography, with a view to significantly advance research. The character of the series is international and multidisciplinary and will include research areas such as: GIS/cartography, remote sensing, geographical education, geospatial analysis, techniques and modeling, landscape/regional and urban planning, economic geography, housing and the built environment, and quantitative geography. Volumes in this series may analyze past, present and/or future trends, as well as their determinants and consequences. Both solicited and unsolicited manuscripts are considered for publication in this series.

SpringerBriefs in Geography will be of interest to a wide range of individuals with interests in physical, environmental and human geography as well as for researchers from allied disciplines.

More information about this series at http://www.springer.com/series/10050

Jari Pohjola · Jari Turunen
Tarmo Lipping · Anna Sivula
Marko Marila

Historical Perspectives to Postglacial Uplift

Case Studies from the Lower Satakunta Region

 Springer

Jari Pohjola🆔
Tampere University of Technology
Pori, Finland

Jari Turunen🆔
Tampere University of Technology
Pori, Finland

Tarmo Lipping🆔
Tampere University of Technology
Pori, Finland

Anna Sivula
University of Turku
Pori, Finland

Marko Marila
Department of Cultures
University of Helsinki
Helsinki, Etelä-Suomi, Finland

ISSN 2211-4165 ISSN 2211-4173 (electronic)
SpringerBriefs in Geography
ISBN 978-3-030-00969-4 ISBN 978-3-030-00970-0 (eBook)
https://doi.org/10.1007/978-3-030-00970-0

Library of Congress Control Number: 2018955448

This Springer imprint is published by the registered company Springer Nature Switzerland AG
The registered company address is: Gewerbestrasse 11, 6330 Cham, Switzerland

Contents

Chapter 1
Introduction

Abstract Postglacial land uplift has shaped the coastline of western Finland during past millennia. The first recordings of the consequences of the shoreline displacement due to land uplift date from the end of the 14th century. Current land uplift is mostly the result of the Weichselian glaciation, which ended about twelve thousand years ago, however, it is well possible that the process is also affected by the earlier glaciations of major ice ages having taken place up to hundreds of millions of years ago. In Finnish archaeology, coastline displacement has been used in dating the prehistoric settlements, as it can be assumed that these settlements follow the shorelines and waterways. By reconstructing the shoreline displacement and comparing the results with radiocarbon dating of findings from prehistoric settlements, new insights can be obtained from both the archaeology and the land uplift modelling point of view. In this chapter the geological and archaeological history of Satakunta is introduced by the example of four prehistoric sites in Lower Satakunta: Kolmhaara, Kuninkaanhauta, Luistari and Tyttöpuisto.

Keywords Land uplift · Glacial periods · Archaeology · Geology · Stone age
Bronze Age · Iron Age

The Weichselian glaciation with its last glacial maximum occurring at about 20 000 BP (years Before Present) caused Fennoscandia to be covered by an ice sheet of approximate maximum thickness of 3500 m. The ice sheet was spread out from Central Europe to the proximity of the Ural mountains in Russia. It is easy to imagine that an ice load of that magnitude exerts a tremendous force that will press the earth's crust downwards. At the beginning of Holocene, the ice sheet started to melt. After a few recession and expansion periods when the Younger Dryas period ended approximately 11 000 BP, the ice sheet withdrew towards the northwest of Fennoscandia. The area became soon covered by vegetation and inhabited by animals as well as hunter-gatherers. At the same time, the depressed crust, when freed from ice, started to rebound.

The Holocene land uplift in the coastal areas of the Northern Baltic Sea has been known for centuries. Recorded land uplift history in Fennoscandia can be dated back to 1491, when the channel from the Baltic Sea to the city of Östhammar in Sweden, became too shallow to maintain maritime transport. The nearby city of Öregrund was chosen to be the new harbour area and the residents of Östhammar moved to Öregrund, which also inherited the city privileges from Östhammar (Ekman 1991; Hárlen 2003). A Swedish professor of astronomy, Anders Celsius, measured the land uplift in 1743 and noticed that the shoreline withdrawal is 13 mm per year near the city of Gävle, but the reason of the water withdrawal remained unknown (Mörner 1979; Ekman 1991; Beckman 2001). In late 1800s the Swedish geologist Gerard de Geer was able to show the relation between ice age, ice recession, land uplift in Fennoscandia and land depression in southern Sweden (Bailey 1943; Mörner 1979; Ekman 1991; Nordlund 2001).

Nowadays isostatic land uplift can be observed annually on the shores of Vaasa-Umeå strait (Poutanen and Steffen 2015). The combined area of 'Höga Kusten' (High Coast) in Sweden, near Umeå and the Kvarken archipelago in Finland near Vaasa, has been induced into the UNESCO World Heritage list due to its unique properties that are caused by the glacio-isostatic land uplift (WHC 2012, 2015). Isostatic uplift has changed the landscape of affected areas significantly. 'Höga Kusten' is perhaps one of the best-known places where land uplift has lifted the top of the local rock hills to over 285 m elevation from the Baltic Sea level since the beginning of the deglaciation of continental ice (Wikipedia 2018). Uplift changes the coastline of the Baltic Sea and this may cause prehistorical artefacts to be found at strange locations when looking from the modern era viewpoint. One such example is Kuninkaanhauta (King's Grave) in Panelia, Eura, which is the largest known cairn in Finland and dated to 3450–3250 BP. It is located in the middle of a farming area at a slowly descending hillside. If the location of the Kuninkaanhauta is assessed from the viewpoint of the current landscape, it makes no sense to bury a leader in a presumably high position in the middle of a meadow or a field, but when putting the burial site into historical perspective, the Kuninkaanhauta would originally be in a beautiful place near the shore of the Baltic Sea. Therefore, from the point of view of interpreting historical facts it is essential that historical sites, located in land uplift areas, will be connected to proper land uplift models in order to examine their location with respect to coastline, lakes, rivers and other landscape features.

The Lower Satakunta region is known for its rich cultural history and several archaeological remains. In this article we use the few radiocarbon dated archaeological sites of the Eura region as microhistorical examples of the need to reconstruct properly the location of the prehistorical shorelines. Unfortunately the Kuninkaanhauta is not one of those radiocarbon dated sites, because there were no human remains left in the cairn. It reminds us of the fact, that we do not even know, if it ever actually was a king's grave.

The location of Lower Satakunta region in Finland is presented in Fig. 1.1. The study concentrates on the region of municipality of Eura and especially on four sites: Kuninkaanhauta, Kolmhaara, Luistari and Tyttöpuisto (Girl's park). A general view of shore level displacement in Satakunta at present day, 4000 BP and 8000 BP, is shown in Fig. 1.2. Throughout this Brief, we use 1950 CE (Common Era) as the value for the present day.

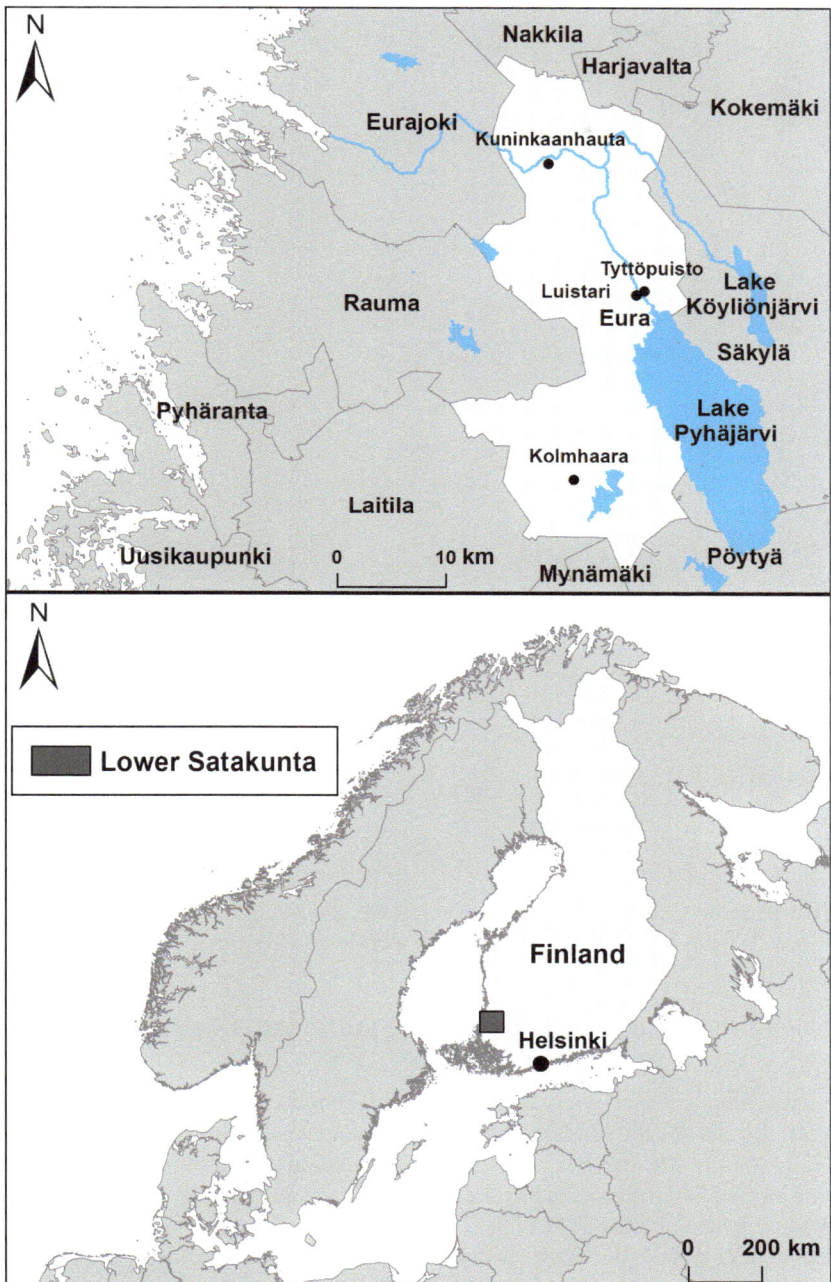

Fig. 1.1 The location of Lower Satakunta region in Finland. The points in the upper subfigure mark the most notable prehistoric sites in Eura. *Background map in upper subfigure: National Land Survey of Finland. Lake and river network data in upper subfigure: Finnish Environment Institute. Background map in lower subfigure: Esri*

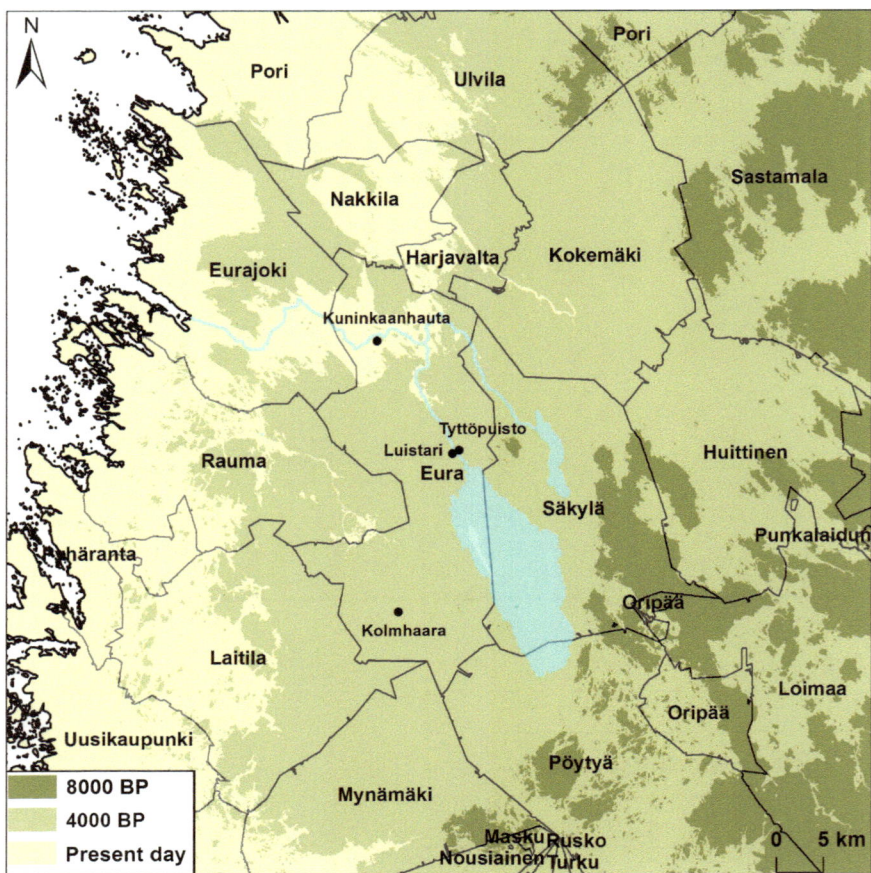

Fig. 1.2 Shoreline location in Satakunta region at 8000 BP, 4000 BP and present day. *Background map: National Land Survey of Finland. Lake and river network data: Finnish Environment Institute*

1.1   Prehistory of Eura Region in Satakunta, Finland

The Mesolithic pioneer population of Eastern Fennoscandia originated mostly in northern Russia and the southeastern Baltic (Rankama and Kankaanpää 2011; Saipio 2017, 242–243). The earliest indications of human habitation in the Eura and inner regions of Satakunta are artefacts related to small human groups of hunter-gatherers, representing the Suomusjärvi Culture. The presence of human populations in the area of Satakunta can be dated back to several centuries before 6950 BP (Huurre 1991, 113–145). Especially the flat chisels and stone clubs prove the area was populated already during the Late Mesolithic period (8450–7050 BP) (Huurre 1991, 121).

In the dawn of the Neolithic period, the landscape of the current area of Eura was a dense archipelago. It provided a favorable environment to the scarce population of seal and beaver hunters and small fisher communities. The beginning of the Neolithic era is, in eastern Fennoscandia, defined by the appearance of pottery. The art of pottery reached this cultural region, through the influences of the upper Volga area, around 7050 BP, forming the basis of the Comb Ceramic tradition (Rankama and Kankaanpää 2011; Saipio 2017, 242–243). The pottery evolved slowly during the 7th and 6th millenium BP (Huurre 1991, 146–147) (Edgren 1984, 24). The Early Comb Ceramic (–5250 BP) was succeeded by the Typical Comb Ceramic (5250–4750 BP). It seems that immigrants from the southeastern lake area brought the typical comb ceramic pottery style to western Finland (Huurre 1991, 188). The Jäkärlä type of ceramics, resembling the early comb ceramics, represents the continuity of the population of the area (Huurre 1991, 302). The Late Comb Ceramics are dated to 4750–4150 BP. The latest period of comb ceramic, the Pyheensilta ceramics prolongs some features of the comb ceramics to 4150–3850/3700 BP. Thus, the tradition of comb ceramics survived from the early Neolithic period to the first centuries of the Bronze Age. Finnish prehistoric periods are presented in Table 1.1.

In the small populations of the Early Neolithic (7050–5950 BP) period in western Finland, agriculture was not a source of living. During the Stone Age, the human groups of the area did not practice large scale cereal cultivation or animal husbandry. Despite the first steps of agriculture taken during the 5th millennium BP, these people would have been highly dependent on hunting and gathering as well as marine food sources. As much as 95% of coastal Stone Age subsistence could have consisted of fish. Until the Early Metal Period (3850–2250/2350 BP), fishing, hunting and gathering were the most important sources of livelihood (Saipio 2017, 243).

The name of the Comb Ceramic culture is derived from the most common type of decoration on its ceramics, which looks like the imprints of a comb. The decorations of comb ceramics took sometimes a form of swimming birds, like in one of the fragments found in Kolmhaara, Eura (Huurre 1991, 169; Äyräpää 1953; Edgren 1966, 8–9). According to the evidence found in Kolmhaara and its surroundings, the

Table 1.1 Finnish prehistoric chronology (according to Saipio 2017, 243)

Name	Period
Early Mesolithic	10850–8450 BP
Late Mesolithic	8450–7050 BP
Early Neolithic	7050–5950 BP
Middle Neolithic	5950–4350 BP
Late Neolithic	4350–3850 BP
Early Metal Period	3850–2250/2350 BP
Early Bronze Age	3550–3050 BP
Late Bronze Age	3050–2450 BP
Early Iron Age	2450 –2250/2350 BP
Middle Iron Age	2350–1150 BP
Late Iron Age	1150–950 BP

people, who were familiar with the art of typical comb ceramic pottery, had a certain sense of luxuries and knew how to wear precious amber pendants. During the period of Typical Comb Ceramic, the deceased of Kolmhaara were buried with their jewels and other grave goods, and covered with red ochre.

The era of Typical Comb Ceramic was an era of population growth in Satakunta. During the period, several new dwelling places occurred, most of them close to the water (Huurre 1991, 169). The area of Eura is rich with the evidence of the culture of typical comb ceramics, but also of the less sophisticated Jäkärlä type of ceramics (Edgren 1966; Huurre 1991, 195). It has been suggested, that during the period of Typical Comb Ceramic culture, there were two different ethnic groups living very close each other, even sharing the dwelling places (Meinander 1984, 36; Huurre 1991, 190–191). We do not know how the cultural interaction was conducted by these different ethnic groups. One group perhaps dwelled on the site in summer and another in winter. One group lived from fishing, the others from hunting. Perhaps they lived side by side, but still maintained the original features of their distinctive cultures.

The spreading of a pottery style does not always indicate much ethnic changes of the populations. The new pottery style may have spread by marriages, or by trade. Objects have been copied. There is, however, evidence of that during the peak of typical comb ceramic, the region of Lower Satakunta had vivid cultural connections to east, west and south.

There is still no consensus as to when cereal cultivation started in the southwestern Finland. Some have suggested that cultivation started as early as 5150–4250 cal BP (calibrated dating Before Present), in the Neolithic Corded Ware culture. Alenius et al. (2009) and Mökkönen (2010) suggest that the first cultivation might have begun even earlier (see also Huurre 2003).

The human carriers of the Corded Ware culture, sometimes referred to as the Battle Axe culture reached the area of Eura between 4750 and 4450 BP. In case of the Battle Axe culture it is even possible to talk about invasion, but the nature of this invasion was perhaps not as much military as cultural. The origin of the newcomers was in the eastern Baltic area. They came across the Gulf of Finland, and groups of them followed the shoreline up northwest. What they found in Lower Satakunta, was the black diabase, a stone suitable for the material of their traditional axes. The corded ware was a novelty when it arrived. It was very different from the earlier pottery. After an interregnum, the population of newcomers began to mix with the already present population. The Corded Ware Culture left permanent marks in the material heritage of the region. Among the new kind of battle and working axes and burial rituals, the newcomers are supposed to have enriched the region of Lower Satakunta with the arts and crafts of agriculture (Huurre 1991, 200). According to many 20th century popular and public interpretations of the prehistory of Finland, the farming was introduced into southern Finland by people of the Corded Ware culture, associated with European agriculture. This interpretation is challenged in current discussions.

Most of the arguments supporting the claim for cultivation in Finland involve pollen as evidence (Mökkönen 2010). Mökkönen used new data concerning changes in settlement pattern, and a correlation between spruce and the Corded Ware culture, to claim that the Corded Ware people were cultivators. The change in settlement pattern was from inland towards coastal sites. According to Lahtinen and Rowley-Conwy, there is, however, no clear reason why such a settlement shift should be connected to cultivation. The settlement shift might as well indicate a shift to a more intensive fishing as an adaptation to climate change (Lahtinen and Rowley-Conwy 2013).

In Finland, the battle axes and corded ware ceramics are found on a quite restricted area. The culture seems not to have had connections to the west, because there are no battle axes of Finnish type found in Sweden. In Finland, the battle axe and corded ware findings are located near the southern and southwestern coastline. The culture of newcomers lived at first isolated from the original population and neither did it have connections to the west. The dwelling sites were not always bound to the shoreline. They could be located several kilometers inland. In Satakunta, most of the battle axes have been found in the south western, lower part of the province. More than 50 battle axes have been found in the area of Eura. It is obvious that the region of Eura, together with the neighboring areas of Kokemäki, Säkylä and Köyliö and part of Rauma, was a provincial center of battle axe culture. The corded ware ceramics are in Satakunta usually found in burial places (Huurre 1991, 212–215).

The makers of these battle axes were skilled professionals, and the presence of the diabase of Eura was a reason why the stone sculptors of the Battle Axe/Corded Ware culture settled to the region. The axes they sculpted were not exported. The production of these artefacts was an industry for the internal market only. The artefacts are supposed to have had a symbolic meaning and been important to the identity of their owners. The traditional pottery style of this distinguished culture survived for centuries side by side with the comb ceramic style, but in the end, when these two different populations and cultures began to merge, evolved to a mixed style of pottery. A collection of battle axes found in the Satakunta region can be seen in Fig. 1.3.

The Battle Axe culture of the shores of the Gulf of Bothnia was a mixed version of cattle-breeder and hunter-gatherer culture (Malmer 1962). As a result of the inter-cultural communication and interaction, whatever it was like, between the original population of the area and the newcomers, a new culture emerged.

The Kiukainen culture is named after the rich dwelling site of Uotinmäki, in The Kiukainen, Eura. The Kiukainen culture was the last Stone Age culture of the south-western coast of Finland, dating to 3950–3450/3250 BP. According to Unto Salo, it was during the period of Kiukainen culture, when the regular practice of agriculture took root permanently to the soils of the southwestern coasts and riverbanks of southwestern Finland (Salo 2000, 75). The Kiukainen culture left behind a practical set of more or less agricultural tools: Working axes, hoes, flint sickles and chisels. When it comes to the material culture, the Kiukainen culture style of pottery was a combination of the heritages of Comb Ceramic culture and the new elements and skills of the Corded Ware culture. In Satakunta, the area of Kiukainen culture was restricted to the coastal area between Eurajoki and Noormarkku. According to Unto Salo, the dwellings of Kiukainen culture continued to be in use during the Bronze Age, and some of them still existed in the dawn of the Iron Age (Salo 1981, 58–61).

Fig. 1.3 Pile of battle axes from Satakunta region, in the exhibition of Satakunta Museum. (Photo: Anna Sivula)

There have been various critiques of these early farming claims. According to Zvelebil and Rowley-Conwy (1984) 'the palynological evidence for Corded Ware farming in Finland rests on a few uncertain identifications of cereal pollen evidence which could hardly justify the belief that the Corded Ware economy was based on farming' (Zvelebil and Rowley-Conwy 1984, 115). Direct evidence continues to elude archaeologists.

It is also possible that animal herding and cereal cultivation were adopted at different times. There is one domestic animal bone, a sheep/goat carpal, that can be connected to the Kiukainen Culture. The bone is from Pedersöre Kvarnabba, remarkably far north on Finland's west coast and is dated to 3679±33 BP (Bläuer and Kantanen 2013).

1.1.1   The Bronze Age

There has recently been a discussion on, how intense the agriculture actually was in the west coast of Finland during the Bronze Age. Previous discussions have alluded to Corded Ware cultivation and considered the Bronze age as a period of intensification of agriculture. Núñez (2004) emphasizes the weaknesses in the evidence, but argues that the Corded Ware culture had been partially agricultural, mainly based on the fact that those people were agriculturalists elsewhere. He believes that the farming of Corded Ware people may in this far north have suffered a setback and disappeared almost totally quite soon. Edgren (1999) in his turn, argues that it is unlikely that cultivation occurred during the Late Neolithic period. Positive indicators of cultivation are lacking. According to Edgren (1999), the cultivation in southern Finland did not intensify until the Pre-Roman Iron Age.

During the Bronze Age, the new wave of newcomers, from the other side of the Gulf of Bothnia, introduced some new technologies of grain cultivation and some agricultural tools, together with the new material of bronze. The introduction of bronze was not a revolutionary event: The amount of bronze artefact finds in Finland totals to only 184 (Soikkeli-Jalonen 2016). Nevertheless, the introduction of bronze can directly be connected to the establishment of new trade routes between Western Finland and Scandinavia.

Via the new cultural connections, the monumental cairns and other new cultural practices and rituals arrived to Lower Satakunta. On the Finnish southwest coast, the cairns containing burnt bone material appear roughly contemporarily with the first Scandinavian bronzes, indicating the extensive influence of the Nordic Bronze Age in the region. Altogether, there have been more than 1900 cairns in the Satakunta region, like cultural footprints of western newcomers in the landscape (Salo et al. 1992; Salo 1981, 206–233).

In the last analysis, we do not know how new and odd these cultural elements of the Nordic Bronze Age actually looked like in the eyes of the people in Eura. It was not an amazing invention to scatter the human bones in cairns, if there already was a prevailing custom of scattering the bones of the deceased relatives around the dwellings of the living (Saipio 2017, 268). In the site of Kolmhaara, the dwelling continued during the Bronze Age, perhaps even to the Pre-Roman Iron Age. The deceased of Kolmhaara were buried near the house they had lived in. The people in Kolmhaara did not adopt the burial rituals of the coastal cairn builders, but treated their deceased relatives in their own way (Salo 1981, 101–102).

The Nordic Bronze Age influences spread from the Finnish southwest coast to the wider area of coastal Finland. Roughly, the inlands of Eastern Fennoscandia were more oriented towards the east, while the western coast was more and more oriented to the west. The archaeological recording of cairns enlightens, however, a much more heterogeneous reality, than a sharp division of the coastal cultural sphere with cairns and the inlands without them. There are cairns of Early Metal Period in the Finnish lake district and the Bothnian Bay area (Saipio 2017, 244). It is, therefore, necessary to make the distinction between the coastal cairns and the inland Lapp cairns and the third type of Bothnian Bay cairns. The coastal cairns were most often located near the dwelling sites, whereas the Lapp cairns were located by the lakeshores. Burned human bone is found both in coastal and inland cairns. In most cairns, there are no grave goods. All organic artefacts and almost all human remains have vanished (Saipio 2017, 246). In Satakunta, burned human bone material has been found in eleven Coastal Bronze Age type cairns. The material has been AMS dated (Accelerator Mass Spectrometry) to in between 3691 and 3006 BP. Unlike the Lapp cairns, these eleven Coastal Bronze Age cairns contained no animal bones (Saipio 2017, 260).

Kuninkaanhauta in Panelia is a monumental cairn, located near the ancient shore of the ancient bay of Panelia, together with more than twenty other cairns (Salo 1981, 135). According to Unto Salo, Kuninkaanhauta probably has had nothing to do with royalties: It is an ancient family grave of a wealthy peasant house. These cairns were not built only for the dead but as well for the living. According to Salo, the size mattered. There was a constant competition between the families on the size of the cairns. The families were bragging with their cairns. The bigger the monument, the wealthier was the house (Salo 1981, 149).

The Bay of Panelia and its neighboring areas have been regarded as the central area of Finland's western Bronze Age culture (Salo 1981, 331–383). Besides the cairns, there have been the dwellings (Harjula 2000, 84). According to Janne Harjula (Harjula 2000) only about half of the original burial cairns in the region of the Ancient Bay of Panelia have survived from the beginning of the 19th century to this day. The silhouette of Kuninkaanhauta serves as a symbol of the prehistory of the area (Harjula 2000, 84). The dwellings near Kuninkaanhauta have been dated to Bronze age (Wallenius 1987, 1988).

During the Bronze Age, the proximity of the archaeological sites to land suitable for cultivation increased. Thus, Zvelebil (1981) concluded that the Bronze Age economy was a combination of farming and hunting. Potentially cultivable areas were rich in natural flora and fauna. As Zvelebil and Rowley-Conwy (1984) note, there is an 'autocorrelation between watercourses, sedimentary basins and soil fertility.' Arable land does not, however, become a major focus of the settlement pattern in southwestern Finland until 1500 BP (Lahtinen and Rowley-Conwy 2013).

Macrofossil evidence for Bronze Age or earlier agriculture remains very scarce. One directly dated domestic animal bone, a cow maxillary molar from Nakkila Viikkala, is dated to 3086 ± 30 BP (Hela-1271) (Bläuer and Kantanen 2013), or 3377–3221 cal BP. This is further southwest than the Kiukainen example mentioned above. These two bones remain the only directly dated evidence of domestic animals in Western Finland that necessarily predate the Iron Age.

1.1.2   The Iron Age

There are several reasons to doubt the very early presence of agriculture in the eastern region of the Scandinavian Peninsula. Most sites at which early agriculture is claimed have only one grain of cereal pollen. In a major review article, (Behre 2007, 203–219) pointed out that pollen analysis is a useful tool, but is not an absolute method for determining the presence of agriculture. Wild grasses can produce pollen of cereal-type. Consequently, any evidence derived from single pollen grains is always questionable. In the earlier research, there have been many possible sources of false interpretations: In some studies, the layer with early farming evidence is not dated properly and the error limits are not discussed (Lahtinen and Rowley-Conwy 2013).

Behre (2007) highlights that it is impossible to prove whether single pollen grains are local or even from cultivated plants at all. A single pollen grain neither proves early cultivation nor the lack of it. There is currently no method to distinguish every single wild grass from cultivated species. It is important to know the local history of the sampling site. According to Lahtinen and Rowley-Conwy (2013), long-distance transportation is always a problem in pollen studies. Pollen can travel not just hundreds of kilometers, but even thousands.

We do not know when the first experiments of cereal cultivation were made in Eura. It is unlikely that they were a success. Prehistoric agriculture was a marginal economic activity so far north, and the evidence of it is rare. Lahtinen and Rowley-Conwy (2013) have argued that the cultivation may have in western Finland begun as late as the start of the Iron Age (c. 2500 cal BP). They do, however, agree with the suggestion, that the Late Neolithic Corded Ware Culture or the Kiukainen Culture perhaps adopted animal herding, but not cereal cultivation. The land uplift had brought fertile lands to the disposal of these cultures.

In the end of the Bronze Age, around 2500 BP, the new raw material of iron and the technologies related to it, were introduced in the eastern parts of the Peninsula of Scandinavia. Before the beginning of the first millennium AD, iron axes, knives and scythes had already been several centuries a vehicle of a revolutionary change of the agriculture of the new era. According to Lahtinen and Rowley-Conwy (2013), there is not enough reliable evidence to support the claim of agriculture in Finland before the Iron Age.

A decline in Picea (spruce) has often been used as evidence for cultivation, because the occurrence of single cereal-type pollen grains, other cultural indicator species, high levels of charcoal and the spruce decline often appear together in pollen diagrams. The explanation offered for this is that slash-and-burn cultivation (Finnish: kaski) worked well in spruce forests (Vuorela 1986). However, in Finland, slash-and-burn -cultivation has been practiced until recently, and the most suitable forest for it consists of relatively young birches and alders (Heikinheimo 1915). A connection between spruce and cultivation is therefore not accepted by all. Rowley-Conwy (1983) argued that a spruce decline can be explained without connection to human activity. Huhtakaski, the traditional northern slash-and-burn cultivation of spruce forests, could therefore be an application of the kaski technique, dating to the historical era.

There are still several unanswered questions, when it comes to the early agriculture in the eastern region of the Peninsula of Scandinavia. Lahtinen and Rowley-Convy give an example: There is no evidence that slash-and-burn cultivation was practiced in the early stages of cultivation. Furthermore, it was probably never important in the western part of the country. Thus, the spread of agriculture into eastern Finland might have been affected by the late innovation of fast-cycle slash-and-burn cultivation. However, more studies are needed from the western part of the country (Lahtinen and Rowley-Conwy 2013). According to Tiitinen (2011) the more precise knowledge we have of the paleo environmental landscapes of these dwellings and other archaeological landmarks, the better we understand the cultures behind them. This is where the advanced land uplift modelling becomes useful for those who take part in the meaning-giving process of cultural heritages.

1.2   Geological History of Satakunta

The bedrock of southwestern Finland was formed 1900–1800 Ma ago, when the Svecofennian island arc complex collided with the over 2500 Ma old archaic continent
in northern and eastern Finland (Pajunen and Wennerström 2010; Rämö and Haapala
2005). The Svecofennian orogeny, or the formation of the Svecofennian mountain
range, was a result of a complex and multiphase process containing several com-
pression and extension phases. During the extension phases the crust thinned and
the formation of the sediment basins started. Magmatic and volcanic activity was
present in this phase. During the collision phases, the crust thickened (Pajunen and
Wennerström 2010).

After the formation of the Svecofennian mountain range, the bedrock cooled down
rapidly and the crust was moving towards isostatic equilibrium. This movement
caused massive tectonic displacements and together with erosion, granitic intrusions
and diabase dikes penetrated into the crust at 1800 Ma. Orogenic collapse is inter-
preted to have happened during 1790–1770 Ma, which stabilized the crust to the
cratonic (stable) stage (Lahtinen et al. 2005).

The next event was the emplacement of rapakivi granites in 1650–1550 Ma, when
the Svecofennian mountain range had already eroded down. The rapakivi plutons of
Eurajoki, Peipohja and Reposaari are estimated to be formed in 1590–1540 Ma,
Rapakivi plutons are presented in pink ovals (Peipohja pluton is above Kokemäki) in
Fig. 1.4. The basins filled with jotnian sediment stones (sandstones) in 1400–1300 Ma
and postjotnian olivine diabases were settled down in 1270–1250 Ma (Pajunen and
Wennerström 2010; Rämö and Haapala 2005). It is also suggested that the formation
of sandstones in Satakunta may have occurred during a much longer period between
1650 and 1250 Ma (Pokki et al. 2013), see for example 'silicate sandstones' in
Fig. 1.4.

Various ice ages have shaped the outer layer of the crust together with erosion.
The massive weight of the ice has also pressed the crust down producing massive
cracks and forcing the crust to find an equilibrium between seawater, crust pressure
from inside the earth, and the ice load.

There have been at least six major glaciation periods in Earth's history that
have lasted several million years: Pongola, Huronian, Cryogenian, Andean-Saharan,
Karoo and Quaternary (Kopp et al. 2005; Sleep and Hessler 2006; Tang and Chen
2013; Plumb 1991; Love et al. 2009; Marshall 2010; Berner 1999; Gibbard and Van
Kolfschoten 2005). The time periods of these glaciation periods are presented in
Table 1.2.

The quaternary period is referred to as an ice age, because at least one continental
glacier, the Antarctic ice sheet, has existed continuously. The quaternary period can
be divided into at least five smaller glacial periods or stadials (Gibbard and Van
Kolfschoten 2005; Ehlers and Gibbard 2008; Kukla 2005), based on Marine Isotope
Cycles (MIS) (Shackleton et al. 2003) as presented in Table 1.3.

The ice load has its effect on the earth's crust; the crust responds to the load
in the following way (Fig. 1.5). The ice load presses the crust down and seawater

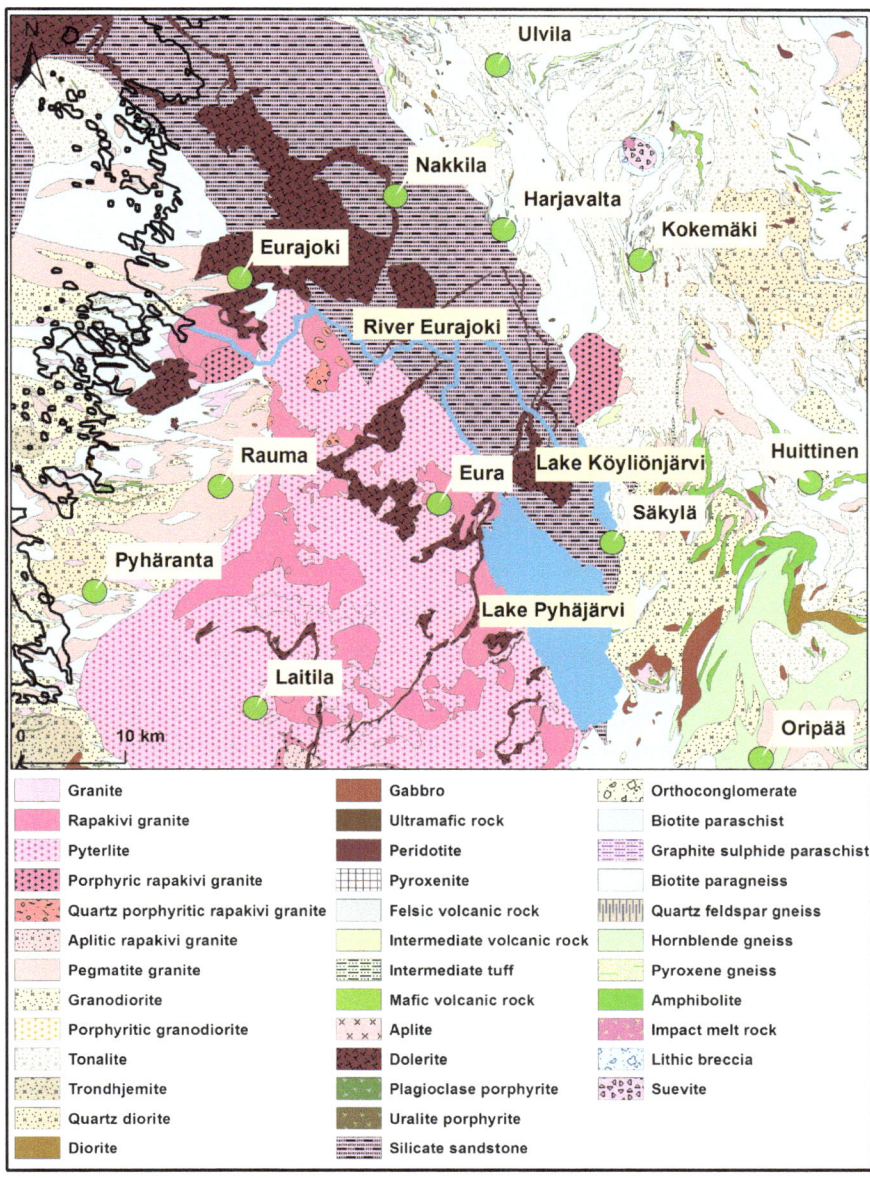

Fig. 1.4 Geological map of Satakunta: Geological Survey of Finland. Lake and river network data: Finnish Environment Institute

Table 1.2 Timeline of major glaciation periods

Name	Period (Ma)
Quaternary	2.588–present
Karoo	360–260
Andean-Saharan	450–420
Cryogenian	720–635
Huronian	2450–2100
Pongola	2900–2780

Table 1.3 Approximate Quaternary glacial periods (stadials) based on MIS dating in northern Europe

Name	Period (Ka)	MIS
Weichselian	110–12	2–4 and 5a–d
Saalian	300–130	6–10
Elsterian	460–380	11–13
Menapian	1400–1180	34–45
Tiglian C4c (pre-pastonian stage)	2100–1850	65–74

gets rearranged according to its gravitational equilibrium. The response of the crust behaves according to the visco-elasto-plastic properties of the crust. After melting of the ice, the melting water mixes with the seawater increasing the seawater volume, thus seeking for a new equilibrium state, and the land will start to uplift as an elastic and viscous response. The elastic response is considered to be fast (Påsse 2001), and after the elastic response, the fast uplift is followed by a slow viscous response. Påsse concluded the possibility of fast uplift by examining the fast recovery of crustal subsidence in the Bothnian Bay region (Påsse 2001).

It is postulated in (Rabassa and Ponce 2016) that during the Middle and Late Pleistocene period the smaller glacial periods called stadials occur in 100,000 years cycles. If this is true with the older major ice ages in Table 1.2, it means that Satakunta may have encountered more than 20 stadial periods. However, the dynamical loads, weights and cumulative pressure effects of pre-Weichselian glaciations are only a guesswork in the sense of Fennoscandian uplift estimation. The possible residual uplift caused by pre-Weichselian glaciations can only be included to be a part of current Holocene uplift.

1.3 Outline of the Brief

In this brief, we use a modified version of Påsse's semi-empirical land uplift model to reconstruct the landscape features in Western Finland, Lower Satakunta. These reconstructions are then used as a basis when examining the rich historical data

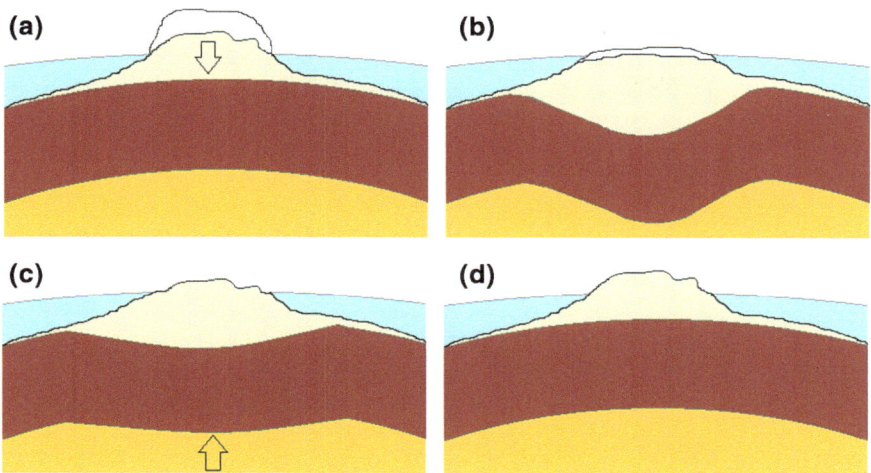

Fig. 1.5 a Ice load will cause earth's crust to respond to the load. **b** After the ice has melted, the earth's crust's uplift process begins first rapidly **c** and then decaying **d** when the crust nears the equilibrium

available from the region by visualizing prehistoric sites such as the Kuninkaanhauta, for example, on top of the reconstructed landscape. This kind of modelling raises several issues related to the exact dating of historical objects, location of historical sites or interpretation of archaeological findings. On the other hand, historical facts help to solve local anomalies and discrepancies in the data used for land uplift modelling and therefore lead to reassessment of the models. Furthermore, these refined models will enhance the reliability of the predictions of landscape development and shoreline displacement during next millenia when planning activities that have long-term effects such as building repositories for spent nuclear fuel, for example.

This brief is organized as follows. In Chap. 2 a detailed overview on cultural history of the Eura area in Lower Satakunta is given. The methods used for land uplift modelling and landscape reconstruction are introduced in Chap. 3. Also the source data underlying the land uplift model are presented in Chap. 3. In Chap. 4, the results of landscape reconstruction are presented together with the visualization of historical sites. Finally, in Chap. 5, conclusions are drawn from the landscape reconstruction.

References

Alenius T, Lavento M, Saarnisto M (2009) Pollen-analytical results from lake katajajarvi–aspects of the history of settlement in the finish Inland regions. Acta Borealia 26(2):136–155. https://doi. org/10.1080/08003830903372043

Äyräpää A (1953) Suomen miekanhiontakivet. Kotiseutu 2:86–91

Bailey E (1943) Obituaries: Baron Gerard de Geer. Nature 152(3847):102–103. https://doi.org/10. 1038/152102a0

Berner RA (1999) Atmospheric oxygen over Phanerozoic time. Proc Nat Acad Sci 96(20):10,955– 10,957. https://doi.org/10.1073/pnas.96.20.10955, http://www.pnas.org/cgi

Beckman O (2001) Anders Celsius. Elementa 84(4):174–180, http://www.astro.uu.se/history/ celsius.pdf

Behre KE (2007) Evidence for Mesolithic agriculture in and around central Europe? Veg Hist Archaeobotany 16(2–3):203–219. https://doi.org/10.1007/s00334-006-0081-7

Bläuer A, Kantanen J (2013) Lammas on ollut kotieläimenä Suomessa kivikaudelta lähtien. Lammas ja Vuohi 4:16–17

Edgren T (1966) Jäkärlä-gruppen. En västfinsk kulturgrupp under yngre stenålder. Suomen muinais-muistoyhdistyksen aikakauskirja 64, University of Helsinki

Edgren T (1984) On the economy and subsistence of the Battle-Axe culture in Finland. Fenno-ugri et slavi 4:9–15

Edgren T (1999) Alkavan rautakauden kulttuurikuva. In: Fogelberg P (ed) Pohjolan poluilla. The Finnish Society of Science and Letters, Helsinki, pp 311–333

Ehlers J, Gibbard P (2008) Extent and chronology of Quaternary glaciation. Episodes 31 (2)(2):211– 218. https://doi.org/10.1016/S0277-3791(03)00130-6

Ekman M (1991) A concise history of postglacial land uplift research (from its beginning to 1950). Terra Nova 3(4):358–365. https://doi.org/10.1111/j.1365-3121.1991.tb00163.x

Gibbard P, Van Kolfschoten T (2005) The Pleistocene and Holocene Epochs. https://doi.org/10. 1017/CBO9780511536045.023

Harjula J (2000) Burial Cairns in the Region of the Ancient Bay of Panelia. Fennoscandia archael-ogica XVI(I):83–102

Hárlen H (2003) Sverige från A till Ö: geografisk-historisk uppslagsbok. Kommentus, Stockholm

Heikinheimo O (1915) Kaskiviljelyn vaikutus Suomen metsiin. Acta For Fenn 4:1–264

Huurre M (1991) Satakunnan kivikausi, Satakunnan historia 1. Satakunnan Maakuntaliitto

Huurre M (2003) Viljanviljelyn varhaisvaiheet. In: Rasila V, Jutikkala E, Mäkelä-Alitalo A (eds) Suomen maatalouden historia 1. Perinteisen Maatalouden aika esihistoriasta 1870-luvulle, Suo-malaisen Kirjallisuuden Seura, pp 19–37

Kukla G (2005) Saalian supercycle, Mindel/Riss interglacial and Milankovitch's dating. Quatern Sci Rev 24(14–15):1573–1583. https://doi.org/10.1016/j.quascirev.2004.08.023

Kopp RE, Kirschvink JL, Hilburn IA, Nash CZ (2005) The Paleoproterozoic snowball Earth: A cli-mate disaster triggered by the evolution of oxygenic photosynthesis. Proceedings of the National Academy of Sciences 102(32):11,131–11,136. https://doi.org/10.1073/pnas.0504878102, http:// www.pnas.org/cgi

Lahtinen R, Korja A, Nironen M (2005) Paleoproterozoic tectonic evolution. In: Lehtinen, Martti (Geological museum of the Finnish museum of Natural History UoH, Nurmi P, Rämö, O T (Department of Geology, University of Helsinki F (eds) Precambrian geology of Finland, Vol 14. Elsevier, p 750

Lahtinen M, Rowley-Conwy P (2013) Early farming in Finland: was there cultivation before the Iron Age (500 BC)? Eur J Archaeol 16(4):660–684. 10.1179/1461957113Y.000000000040, http:// www.tandfonline.com

Love GD, Grosjean E, Stalvies C, Fike DA, Grotzinger JP, Bradley AS, Kelly AE, Bhatia M, Meredith W, Snape CE, Bowring SA, Condon DJ, Summons RE (2009) Fossil steroids record the appearance of Demospongiae during the Cryogenian period. Nature 457(7230):718–721. https:// doi.org/10.1038/nature07673

Malmer M (1962) Acta Archaeologica Lundesia. Jungneolitische Studien, Volume 8, issue 2, University of Lund

Marshall M (2010) The history of ice on Earth. https://www.newscientist.com/article/dn18949-the-history-of-ice-on-earth/

Meinander CF (1984) Kivikautemme väestöhistoria–Suomen väestön esihistorialliset juuret. In: Gallén J (ed) Bidrag till kännedom av Finlands natur och folk. Utgivna av Finska Vetenskaps-Societeten, Finska Vetenskaps-Societen, Helsinki, pp 21–49

Mökkönen T (2010) Kivikautinen maanviljely Suomessa. Suomen Museo 2009:5–38

Mörner NA (1979) The Fennoscandian uplift and late cenozoic geodynamics: geological evidence. Geojournal 3(3):287–318. https://doi.org/10.1007/BF00177634

Nordlund C (2001) Det Upphöjda Landet. Vetenskapen, landhöjningsfrågan och kartläggningen av Sveriges förflutna, 18601930. PhD thesis, University of Umeå

Núñez M (2004) All quiet on the Eastern Front? In: Knutsson H (ed) Coast to coast—arrival results and reflections. University of Uppsala, Department of Archaeology and Ancient History, pp 345–367. https://doi.org/10.2307/1148166

Påsse T (2001) An empirical model of glacio-isostatic movements and shore-level displacement in Fennoscandia. Technical Report, Swedish Nuclear Fuel and Waste Management Co, http://www.skb.se/upload/publications/pdf/R-01-41.pdf

Pajunen M, Wennerström M (2010) Satakunnan Hiekkakiven Hauraiden Rakenteiden Kehityksestä. 183, Geological Survey of Finland

Plumb K (1991) New Precambrian time scale. Episodes 14(2):139–140

Pokki J, Kohonen J, Lahtinen R, Rämö OT, Andersen T (2013) Petrology and provenance of the Mesoproterozoic Satakunta formation, SW Finland. 204, Geological Survey of Finland

Poutanen M, Steffen H (2015) Land uplift at Kvarken Archipelago/high coast UNESCO World Heritage area. Geophysica 50(2):49–64

Rämö OT, Haapala I (2005) Chapter 12 Rapakivi Granites. Dev Precambr Geol 14(C):533–562. https://doi.org/10.1016/S0166-2635(05)80013-1,9809069v1

Rankama T, Kankaanpää J (2011) First evidence of eastern Preboreal pioneers in arctic Finland and Norway. Quartar 58:183–209

Rabassa J, Ponce JF (2016) The Heinrich and Dansgaard-Oeschger Climatic Events During Marine Isotopic Stage 3. Marine Isotope Stage 3 in Southern South America, 60 ka BP30 ka BP 3(Mis 3):7–22. https://doi.org/10.1007/978-3-319-40000-6

Rowley-Conwy P (1983) Reviews. J Dan Archaeol 2:205–209

Salo U (1981) Satakunnan historia I,2 : Satakunnan pronssikausi. Satakunnan Maakuntaliitto

Salo U, Tuovinen T, Vuorinen JM (1992) Luettelo Suomen hautaraunioista. Osa 1. Turun ja Porin läänin varhaismetallikautiset hautarauniot. Karhunhammas 14:1–94

Salo U (2000) Ihmisen jäljet Satakunnan maisemassa. Suomalaisen kirjallisuuden seura, Helsinki

Saipio J (2017) The emergence of cremations in Eastern Fennoscandia. Changing uses of fire in ritual contexts. In: Cerezo-Román J, Williams H, Wessman A (eds) Cremation and the archaeology of death. Oxford University Press

Shackleton NJ, Sánchez-Goñi MF, Pailler D, Lancelot Y (2003) Marine isotope substage 5e and the Eemian interglacial. Glob Planet Change 36(3):151–155. https://doi.org/10.1016/S0921-8181(02)00181-9

Sleep NH, Hessler AM (2006) Weathering of quartz as an Archean climatic indicator. Earth Planet Sci Lett 241(3–4):594–602. https://doi.org/10.1016/j.epsl.2005.11.020

Soikkeli-Jalonen A (2016) Pronssikautiset pronssiesineet Suomessa. Pro gradu thesis, University of Turku

Tang H, Chen Y (2013) Global glaciations and atmospheric change at ca. 2.3 Ga. Geosci Front 4(5):583–596. https://doi.org/10.1016/j.gsf.2013.02.003

Tiitinen T (2011) Liikettä ajassa ja paikassa–Lounais-Suomen muinaisrannat tarkastelussa. In: Uotila K (ed) Avauksia Ala-Satakunnan esihistoriaan. Eura Print Oy, Eura, pp 47–80

Vuorela I (1986) Palynological and historical evidence of slash-and-burn cultivation in South Finland. In: Behre KE (ed) Anthropogenic indicators in pollen diagrams. Balkema, Rotterdam, pp 53–64

Wallenius T (1987) Kiukainen panelia kuninkaanhauta. Technical report, Museovirasto

Wallenius T (1988) Kiukainen Panelia Kuninkaanhauta Vanhemman metallikauden asuinpaikan kaivaus. Technical report, Museovirasto

WHC (2012) World heritage cooperation. Kvarken. https://doi.org/10.1108/S0749-742320160000019022, www.kvarkenworldheritage.fi/assets/Svenska-pdf/PPKvarkeneng2.pdf,0712.0689

WHC (2015) World heritage list. https://whc.unesco.org/en/list/898

Wikipedia (2018) High coast. https://en.wikipedia.org/wiki/High_Coast

Zvelebil M (1981) From forager to farmer in the Boreal zone, vol 115. British Archaeological Reports, Oxford

Zvelebil M, Rowley-Conwy P (1984) Transition to farming in Northern Europe: a hunter-gatherer perspective. Norw Archaeolo Rev 17(2):104–130. https://doi.org/10.1080/00293652.1984.9965402, http://www.tandfonline.com/

Chapter 2
Overview of the Prehistory of Eura: Life Around a Few Radiocarbon Dated Spots

Abstract The rich and colorful human history of Eura has been documented based on archaeological findings. The oldest findings from Satakunta are associated with the Suomusjärvi culture, which may have had possible cultural and trading connections to East Karelia. The findings also show that there has been two different types of ceramics in the same place and time period indicating a chance that there have been two populations from different cultures at the same time. Numerous archaeological findings belonging to the later Kiukainen culture show the cultural dependencies with other archaeological finding sites in Finland. In this chapter, four archaeological examples that belong to either Suomusjärvi or Kiukainen culture are presented. Also a brief history of Scandinavian archaeology is presented and the roots of Finnish archaeology are discussed in this chapter.

Keywords Archaeology · Dating · Kiukainen culture · Suomusjärvi culture

Teija Tiitinen has emphasized that the current interpretations of the Stone Age and Bronze Age shorelines in southwestern Finland must be reconsidered. Although the elevation of the site is most often used in dating the dwelling places of Stone Age, they sometimes are used in the dating of the coastal metal cultures of southwestern Finland as well (Tiitinen 2011). The identification of the accurate shoreline is not only a dating tool, but an important tool of the reconstruction and representation of a inhabited landscape.

The paleo-environmental context of an archaeological site has been an important tool of archaeological interpretation since the 1840s, and in Finland the land uplift has been taken into account as a historical fact already as early as 1915. Following the publications by Eronen et al. (1995) and Eronen et al. (2001) on land uplift in Finland, there has been discussion on the need of more precise modelling of the Stone Age and Early Metal Period shorelines. The new modelling methods have revealed that the uplift is slower than presumed earlier. The lived landscapes may have been very different from the earlier assumptions. Therefore the future research must focus on the locations in the landscape, especially on the location of dwellings and their relation to the shoreline (Tiitinen 2011, 47–49).

© The Author(s), under exclusive licence to Springer Nature Switzerland AG 2019
J. Pohjola et al., *Historical Perspectives to Postglacial Uplift*,
SpringerBriefs in Geography, https://doi.org/10.1007/978-3-030-00970-0_2

2.1 The Shores of the Stone Age (–3500 BP) of the Eura Region Revisited

We cannot assign an exact date to the beginning of the population in the region of Eura. There are no datable finds of human remains. Some of the earliest archaeological findings, such as the dwelling site of Kolmhaara, in Honkilahti, have been considered to be remnants of the Suomusjärvi Culture (8500–6200 BP). Around 7000 BP the Eura region belonged to the dense archipelago of the Litorina Sea, and it was already populated. The small populations of Stone Age Western Finland did not practice large scale cereal cultivation or animal husbandry, so they would have been highly dependent on hunting and gathering, especially seal hunting, as well as other marine food sources. As much as 95% of coastal Stone Age subsistence could have consisted of fish. The stone gouges found in Eura and surrounding areas are made of green slate of the area of Lake Ääninen, Russia, which may indicate there has been some early connections to the east. The findings of dark slate objects in the Lähdekorpi, Eura near Lake Pyhäjärvi, indicate some connections to the area of Lapland. Around 4500 BP the area had active connections and immigrants from the south. (Salo 2000; Edgren 1984) The area of Eura is a rich source of findings dated to the stone age. In the current region of Eura more than 1000 stone age artefacts and approximately 40 dwelling sites have been discovered (Huurre 1991, 108).

The site of Kolmhaara, Eura, with the surroundings, is one of the most famous archaeological sites in Finland. It is a wide dwelling site. The spreading of Typical Comb Ware has been traditionally explained with a rapid cultural diffusion or even immigration from the southeast to western Finland. The AMS datings of ceramics do not support this view. The oldest Typical Comb Ware has been found in Kolmhaara and in Northern Ostrobothnia. The calibrated datings are so near to the datings of the Typical Comb Ware in the Carelian Isthmus, that it must be assumed, that the Typical Comb Ware spread very rapidly to its whole distribution area (Mökkönen 2010, 304). The site of Kolmhaara was discovered by Väinö Lintovaara in 1938 (Edgren 1966, 17–51). It is unfortunately partly destroyed. From the 7000 fragments of pottery found in Kolmhaara, $1/4$ is of Jäkärlä type and $3/4$ of Typical Comb Ware type. The fragments of the Jäkärlä Ware have been counted to be belonging to 75 different items. There have been found more than 80 amber artefacts, 26 graves, remains of houses and several stoves. There are actually three different sites in Kolmhaara: The north hill, the south hill and the site of Munasaari. The locations of the sites are presented in Fig. 2.1 and the site Munasaari is illustrated in Fig. 2.2. 800 m west from the Kolmhaara site there is the site of Lammila, a dwelling site with a substantial amount of fragments of Jäkärlä ceramics and 2–3 stoves. Kolmhaara provides evidence of the connections between east and west (Huurre 1991, 175–176; Edgren 1966, 50).

Small findings, such as differences in the structures of fishhooks, the presence of eastern imported stones with the typical comb ceramics and the absence of them with the Jäkärlä type of comb ceramics, may indicate, that there were two different populations in the area, in same time. There are also differences in the forms of shovels and knives, and there are some special types of drills, that are often present

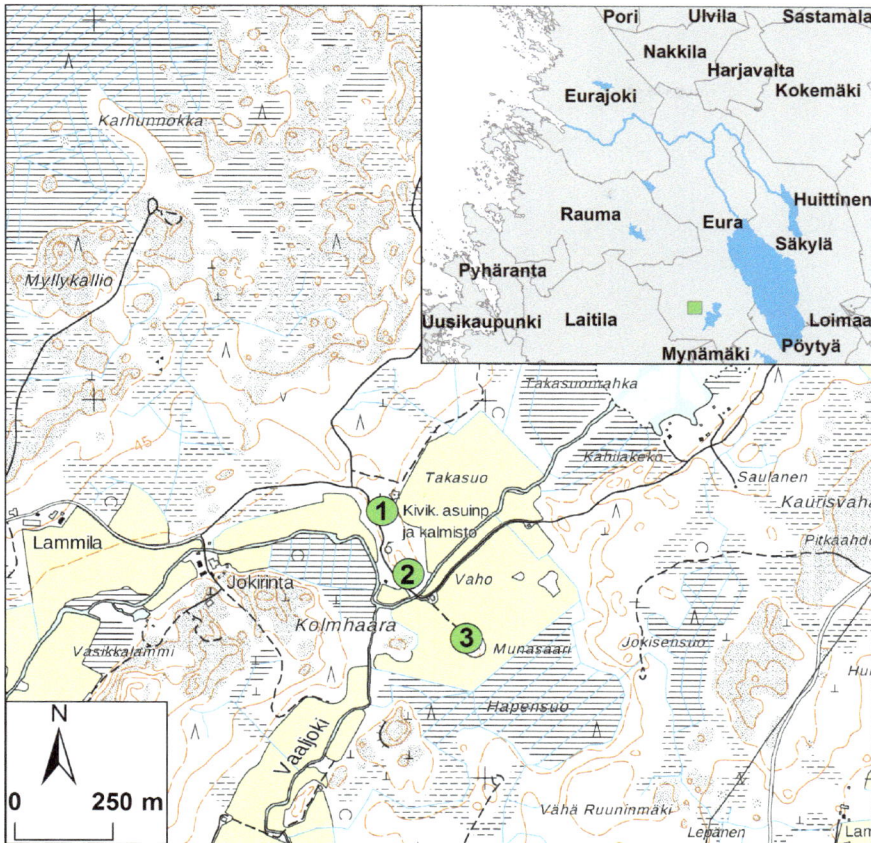

Fig. 2.1 The location of the north hill (1), the south hill (2) and the site of Munasaari (3) in Kolmhaara, Eura. The location of Kolmhaara in WGS 84 coordinate system: 60° 57′ 9,793″, 22° 3′ 49,435″. *Background maps: National Land Survey of Finland. Lake and river network data: Finnish Environment Institute.*

with the Jäkärlä Type of Comb Ceramics but absent from the Typical Comb Ceramics (Edgren 1966, 124–140, 149). There are also different burial types, that indicate that the eastern and western cultural influences have met in the dwelling site of Kolmhaara (Huurre 1991, 182–183). People have lived in the site of Kolmhaara during a period of 600 years, (6212 ± 177 BP–5888 ± 91 BP) The site of Munasaari has been lived and dwelled landscape from 6800 BP to 5400 BP. Altogether this means a 1400 years period of dwelling. There are also some marks of the site being used in the iron age. Teija Tiitinen has pointed out, that there is an anomaly in the dating of the Munasaari site. According to Tiitinen (2011, 76) the height location of the site gives a later date than the AMS dating, if derived from the methodology of Eronen et al. (2001).

Fig. 2.2 Munasaari in Kolmhaara, Eura. The photo has been taken from the south hill's direction. (Photo: Jari Pohjola)

The dwelling site of Tyttöpuisto, Eura, has served as a lived Stone Age landscape during a quite short period. The analyzed findings are all dated to a period of 300 years. The four samples of carbon, from the stoves, are dated between 5819 and 5721 (±110) BP. There are also findings of corded ware in the upper areas of the site and Jäkärlä type of pottery in the lower areas of the site. There seem to have been at least two different dwellings on the site (Tiitinen 2011, 68–69). The dwelling at Kauttua seems to have emerged in the time of the necking of the lake Pyhäjärvi. The Tyttöpuisto site was located by the river, near the rapids. Unlike many other Stone Age dwellings of Eura, Tyttöpuisto has not been situated on an island (Huurre 1991, 164). According to Tiitinen, Tyttöpuisto is an example of a dwelling site, the historical reconstruction on the map is still under discussion, because of the different interpretations based on the different land uplift and sea lowering models. According to Tiitinen, the previous land uplift models are not matching the results of the ^{14}C dating (Tiitinen 2011, 78–79). The location of the site is presented in Fig. 2.3.

The Corded Ware culture, (or Battle-Axe culture), landed on the shore of southwestern Finland between 4800 and 4500 BP. The origins of the new customs, artefacts and traditions were in the Baltic region and in the northeastern region of Poland. The new culture was very different from the earlier culture of southwestern Finland, and a new era began (Haggrén et al. 2015, 83). The period of corded ware (or battle-axe culture) preceded the last Stone Age culture of the southwestern coast of Finland: The Kiukainen culture. The archeological name of the Kiukainen culture can be traced to dwelling sites of Uotinmäki, Kiukainen, in the region of Eura (Räty 1988), but the area of the Kiukainen culture ranged from the shore of Kvarken to Vyborg Bay. In its style of pottery, the Kiukainen culture combined the elements of the earlier Pit-Comb Ware and later Corded Ware cultures. The Kiukainen culture, dating to

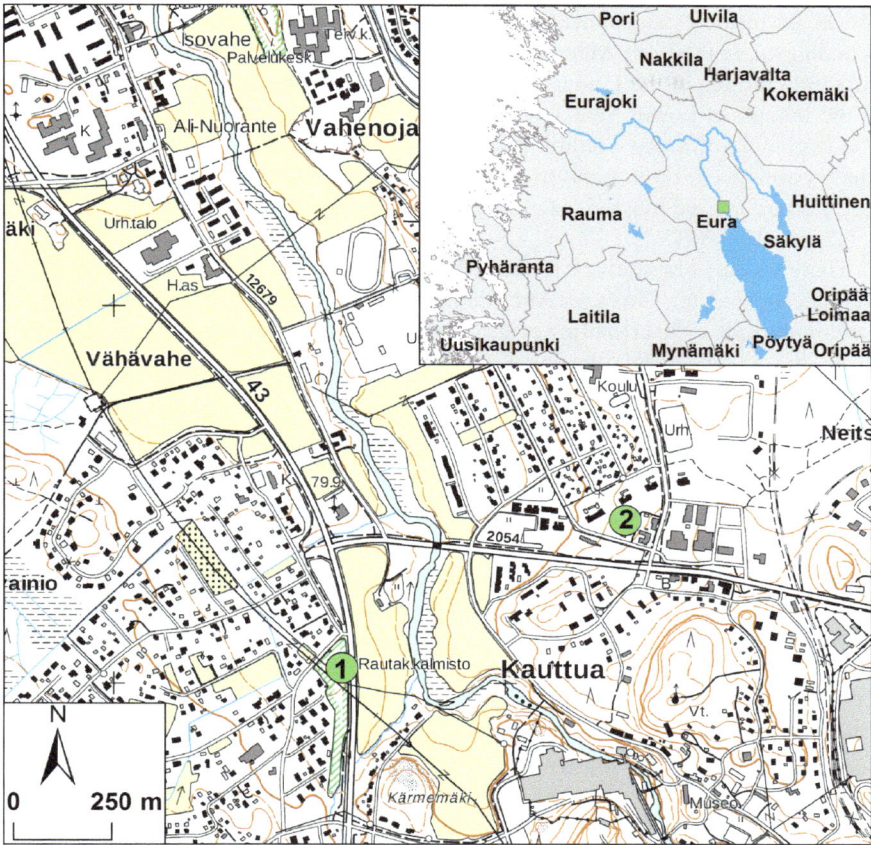

Fig. 2.3 The location of Luistari (1), WGS84: (61° 6′ 45,331″, 22° 8′ 46,374″) and Tyttöpuisto (2), WGS84: (61° 7′ 0,372″, 22° 9′ 33,811″) in Eura. *Background maps: National Land Survey of Finland. Lake and river network data: Finnish Environment Institute.*

4000–3300 BP, emerged within the battle-axe culture and, at the end of the period, it marked the transition from the Stone age to the Bronze age.

2.2   Littoral Dwellings and the Empty Cairns: Bronze Age in Eura (3500–2500 BP)

Reconstructions of Bronze Age habitation and population size in the Eura region are mainly based on the distribution of burial cairns and bronze artefacts as well as changes in ceramic style. The amount of bronze artefact finds in Finland totals to only 184 (Soikkeli-Jalonen 2016). Nevertheless, the use of bronze is directly connected to the establishment of new trade routes between Western Finland and Scandinavia. In the Early Metal Period, especially in the Bronze Age there were two different cultures in the Peninsula of Finland. One in the coastal area and another in the inland. In Lower Satakunta, the early Bronze Age begun around 3500 BP. The beginning of a new cultural period is marked by the stone burial monuments. These cairns were often built on solid rock hills. The cairns of the coastal Bronze Age Culture are the most visible prehistoric monuments in the landscape of Lower Satakunta (Raike 2012).

There are only few artefacts found in the cairns of Satakunta, and therefore the sites are not easy to date. The exact AMS dating of the burned human bone material was not possible until 2001 (Lanting et al. 2001). Because of the possibility of secondary burial, the date gives only the age of the burnt bones, and a point of ex-ante, but not the exact date of the building of the cairn.

The landscape inventory European Pathways to Cultural Landscapes (Clark et al. 2003, 74–83) pointed out that like the dwellings of the western Bronze Culture in Finland, the Stone Age dwelling sites had also been located near the shoreline and, during the Stone Age they seem to follow the changes of the shoreline. Another discovery of Tiitinen (2011) is that the early metal period dwellings and cairns are not always located near to their dated shoreline. For example in Laitila, near Eura, 14% of the bronze age cairns are located more than 500 m from the shoreline that is supposed to have been at a height of 20 m. Either the location of a cairn by the shore was not culturally as preferred as we have believed, or the current dating of the cairns is erroneous, or the modelling of the ancient shorelines should be more precise than it currently is (Tiitinen 2011, 54).

The building of the monumental cairns of the early Bronze Age ended in Satakunta in 2800–2600 BP, but in some cases the cremated remains of later deceased people were buried under the stones of old cairns. This cultural feature is one possible source of the incorrect datings.

The most famous cairn remaining from the era is called the Kuninkaanhauta and it is located in the former municipality of Kiukainen that was consolidated with Eura in 2009. The cairn is located right beside the village road of Panelia. The cairn has a diameter of 35 m and its height is 4–5 m. The location of the site is presented in Fig. 2.4 and the cairn is illustrated in Fig. 2.5.

2.3 The Iron Age

Luistari (Eura) is a multiperiodic archaeological site. It consists of remnants of an Early Metal Period dwelling site with many stoves and cairns, and a cemetery of more than 1300 graves, dating from the 1450 BP to the 850 BP (Pukkila 2011; Salo 1981, 96–100). Multiperiodic sites are interesting, from the point of view of their possible littoral location. It seems, however, that Luistari holds a much longer continuity as a burial plot, than as a dwelling site. According to Nora Kivisalo, the archaeological material from the Luistari cemetery in Eura reveals the increasing role of hunting

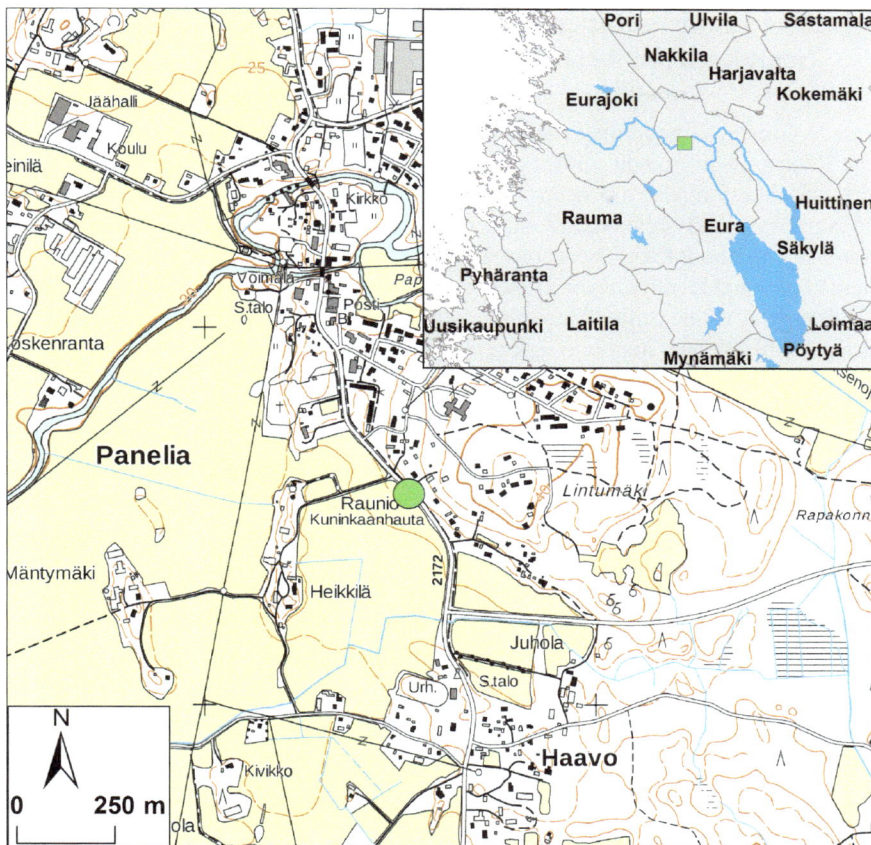

Fig. 2.4 The location of the Kuninkaanhauta in Panelia, Eura. WGS84: (61° 13′ 9,052″, 21° 58′ 46,244″). *Background maps: National Land Survey of Finland. Lake and river network data: Finnish Environment Institute*

in Late Iron Age societies (Kivisalo 2008, 264). Pirkko-Liisa Lehtosalo-Hilander (Lehtosalo-Hilander 2000, 204) has linked this with the increased importance of wilderness utilisation and especially the fur-hunting economy. The fur-based economy strengthened the social position of women in the society of Late Iron Age. On the basis of the archaeological record, in particular the rich female inhumation graves at the Luistari cemetery, Lehtosalo-Hilander has outlined a picture of active and economically powerful women in the Late Iron Age (Lehtosalo-Hilander 2000, 307). The location of the site is presented in Fig. 2.3. In Fig. 2.6 a rock assembly in the site area can be seen.

2.4   The Finnish Stone Age in the Prehistoric Chronology of Scandinavian Archaeology

Archaeology as a modernistic science has its roots in the era of the enlightenment. The Renaissance brought with it a renewed interest in and an appreciation of history, and therefore also a concern for the remains of the antiquity. Early antiquarian archaeology mainly revolved around collecting with no concern for an explicit and systematic methodology. The leading motivation behind collecting was the establishment of national identity beyond that demonstrable through historical documents; the development of scientific methodology was secondary. Mid-19th century, however, brought with it many methodological advances.

One of these methodological advances was the construction of the conceptual foundations of an archaeological chronology (Gräslund 1987). Based on his observations while organising museal collections in the early decades of the 1800s, Danish archaeologist Christian Jürgensen Thomsen reintroduced the then already well-established idea of the Three-Age System (Klindt-Jensen 1975, 50–51). According to this idea, archaeological artefacts could be divided into three groups, each denoting a specific archaeological age, the stone age, the bronze age, and the iron age, respectively. Although material traits played a significant role in this development, the classification reflected museal rather than archaeological realities.

Fig. 2.5 Kuninkaanhauta, Eura. (Photo: Jari Turunen and Jari Pohjola)

Fig. 2.6 Artificial rock assembly in Luistari, Eura. (Photo: Jari Turunen)

The truly significant insight of Thomsen, then, was to emphasise the archaeological context of the finds. While the material of an artefact was no doubt important for Thomsen's chronology, he nevertheless felt that 'nothing is more important than to point out that hitherto we have not paid enough attention to what was found together' (Gräslund 1987, 23). Because stone and bronze artefacts were more often found together than stone and iron artefacts, and because the relatively younger age of iron artefacts could be established with recourse to written sources, Thomsen inferred that stone artefacts must be older than bronze artefacts.

Thomsen's contextual-comparative method was further developed in Danish archaeology, most famously by Sophus Müller, (Müller 1884, 1897), who based his culture-historical archaeology on the idea that the geographical areas inhabited by distinct ethnic groups were marked by the existence of specific find combinations peculiar to those groups. The present day material culture of an ethnic group, in turn, formed the starting point for this comparison. In one sense then, the ethnic group was seen as a historical delineation of a particular group of people that had a distinct geographic origin. Consequently, views similar to Müller's were adopted widely in nationalistic research in Finland by the father of Finnish archaeology Johannes Reinhold Aspelin (Salminen 2013, 26). Aspelin was one of the first Finnish archaeologists to systematically describe and compare Finnish archaeological finds in order to establish the origins of the Finns. By comparing Finnish archaeological finds to those found in neighbouring areas, Aspelin argued that the Finns originated in the Altai region, and migrated to Finland via the Baltic in the eight century AD, prior to which the area had been occupied by Germanic or Scandinavian tribes (Aspelin 1875, 1885).

Another significant conceptual revolution in the construction of the archaeological chronology, then, took place in the wake of Darwin's theory of evolution, published in 1859. While Thomsen's chronological approach was founded on the idea of context, the evolutionary approach came to stress the evolution of the artefact type. Famous proponents of the typological approach include the Swedes Oscar Montelius and Hans Hildebrand. In his Scientific Archaeology, Hildebrand (Hildebrand 1873) writes that [u]nder the influence of two factors—the practical need and the craftsmans taste—a great many forms arise, each of which has to struggle for its existence; one does not find what it need for its existence and succumbs, but the other moves forward and produces a whole series of forms (Hildebrand 1873) translation from (Gräslund 1987, 101).

Some years later, Montelius put forth his famous adage 'what the species is to the natural scientist, the type is to the archaeologist,' elaborating that the task of the archaeologist should no longer be to simply describe and compare the finds and the cultures, but instead, by following the law of evolution, to trace the internal connection which exists between the types, and to show how one type has developed from the other (Montelius 1884, 1899). Although Hildebrand's and Montelius' evolutionary typology stands in great contrast to Müller's ethnohistorical approach, it should be noted that they share a partially common genealogy. Montelius' typological method was clearly informed by the findings of the comparative method. These early examples of the construction of the archaeological chronology have mutually contributed to the development of Scandinavian archaeology in the course of the 19th century, mainly resulting in further and exceedingly finer divisions within the three ages, such as the stone age into the Palaeolithic, the Mesolithic, and the Neolithic period, for instance.

Furthermore, the two methods were often combined effortlessly. German linguist and archaeologist Gustaf Kossinna combined both in his dogmatic settlement archaeology. Like Aspelin, Kossinna identified culture group with ethnic group, and held accordingly that changes in the material culture always reflected diffusion of ideas rather than migration of people (Salminen 2003, 30; Baudou 2005). Similar ideas for the causes behind changes in material culture were also adopted to Marxist Soviet archaeology (Bulkin et al. 1982). It is, however, noteworthy that these forms of German and Soviet archaeology were also based on an evolutionary conceptualisation of cultural change, and included the explicit idea of inherent racial superiority or social stratification.

However, the two approaches also mark two distinct understandings of the nature of the preferred methodology of archaeology. Whereas ethnohistory saw culture as the defining archaeological entity, evolutionary typology saw the human agent as subordinate to the laws of nature. Accordingly the two came to favour different research methods. Both methodological positions were criticised by Finnish archaeologist Aarne Michaël Tallgren in the 1930s (Tallgren 1934, 1937). Although influenced by Montelius, Tallgren never adopted his natural scientific methodology. Tallgren was critical of archaeology's development into a natural science and claimed that it entailed a methodological cul-de-sac. Ideologically, Tallgren was closer to Müller, but rejected Kossinna's settlement archaeology. Instead, Tallgren favoured a social approach to culture: 'I regard culture as a human product and not as a natural product. It is a social product and it should be studied as such' (Tallgren 1937, 159).

The straight identification of Darwinist evolutionary ideas behind the development of the typological method is, however, misleading. Although typological sequencing may find an analogue in some type of evolutionist thinking, Darwinian or pre-Darwinian, it nevertheless was based on observations of stylistic, formal, and material changes in the archaeological artefacts. Interestingly, then, Bo Gräslund, in his The birth of Prehistoric Chronology (Gräslund 1987) provides one reason why the descriptive method may not differ significantly from the typological method. For (Gräslund 1987, 108), the reason concerns the language of science: 'When the archaeologists had to describe typological connections, whether descriptive or actively grading, they did so, linguistically, by consistently allowing the artefact material to appear as the subject instead of the object. The correct procedure would of course have been to represent the makers, the ancient craftsmen, as the active force behind the changes. In this way, it became natural to design the description as if the mechanism of change lay instead in the dead material.'

What has been characteristic of the methodological development of archaeology since its earliest iterations, then, is a constant oscillation and renegotiation between the methods of the natural sciences and the humanities (Marila 2018). In this respect, the methods of the humanities are often seen as simply descriptive, while the task of natural sciences is often seen to provide causal explanations for the witnessed phenomena. In archaeology, the causal explanations are commonly provided in recourse to law-like natural processes, such as the aforementioned Darwinian evolution, or geological stratigraphy.

Before wider acceptance of Darwin's theory, the demonstration of the deep history of man had rested on geology. Up until the mid-19th century, remains of extinct animals but no man-made objects that differed significantly from known stone-age artefacts had been found in Pleistocene layers. If such artefacts were found, the antiquity of man could be established beyond creationist calculations, such as the 4004 BC provided by archbishop Usscher. Coincidentally, in 1859, in the same year that On the Origin of Species was published, a significant archaeological discovery was made in St Acheul, France, where a Palaeolithic stone tool was found in one of these deep Pleistocene layers, see also (Gamble and Kruszynski 2009). In 19th century European prehistoric archaeology, geological stratigraphy came to be methodologically associated with the study of artefact typology, and the two formed the scientific backdrop for the establishment of prehistoric chronology.

Geology has also been important for the development of Finnish archaeology. The first relative chronology for the Finnish Stone Age resulted as a joint effort between geologist Wilhelm Ramsay and archaeologist Aarne Äyräpää in the 1920s (Ramsay 1920, 1926; Europaeus 1923, 1926; Europaeus-Äyräpää 1930) see also (Siiriäinen 1989). The seeds of Äyräpää's chronology were presented in an article published in 1923. Based on pottery fragments found at the Säkkijärvi Ravi site from different stratigraphical layers and different elevations, Äyräpää concluded that some of the finds were older than others. So-called Comb Ware (pottery decorated with comb stamps and pits) was found in deeper layers and at higher elevation than so-called Corded Ware, the type of pottery associated with the late Neolithic Scandinavian Battle Axe culture (see also Europaeus 1926, 47).

Earlier, Finnish stone age scholar Sakari Pälsi (Pälsi 1915) had identified Comb Ware as characteristic of the Finnish Neolithic, and further grouping was done by Julius Ailio (Ailio 1922), who identified two additional groups, the early and the late forms. Pälsi's and Ailio's chronologies were, however, mainly based on functional and stylistic criteria, and lacked chronological rigour. Based on Ramsay's shore displacement chronology of Southern Finland, Äyräpää was able to provide, for the first time, a relative chronology for the Comb Ware (Europaeus 1923, 1926; Europaeus-Äyräpää 1930). Following the grouping proposed by Ailio, Äyräpää divided the Comb Ware pottery into three consecutive groups, Early Comb Ware, Typical Comb Ware, and Late Comb Ware, each style denoting a distinctive elevational sequence (Europaeus-Äyräpää 1930). Above these levels, one would find remains of Mesolithic cultures that were not yet producing pottery.

It should be noted that the use of land uplift in sequencing archaeological finds was, and still is, based on the idea that Stone Age populations were shorebound, and that the dwellings would therefore most likely be located within one to two metres' height difference in relation to the contemporary shoreline. Therefore water levels could only be used to indicate the minimum age of the find. Äyräpää acknowledged this methodological challenge, and chose explicitly to identify particular finds with their actual elevation rather than speculate on their distance to the contemporary shoreline (Europaeus 1923, 30) (see also Meinander 1951).

For Äyräpää, Typical Comb Ware represented the pinnacle of Finnish Neolithic pottery in terms of stylistic and technological refinement, and Typical Comb Ware became the defining term for the Finnish Comb Ware, but also the type through which other types of ceramics were defined (Siiriäinen 1989). Consequently, Äyräpää's chronology was passed on for decades in Finnish archaeology as 'a known fact', which hindered subsequent research on the topic (Nordqvist and Mökkönen 2015). In the course of the 1960s and the 1970s, Äyräpää's chronology became increasingly challenged. While Äyräpää's chronology had utilised regional land-uplift isobases, such as those provided by Ramsay, subsequent chronologies were calculated from a gradient/time curve (Siiriäinen 1969, 1972, 1973). Because of his reliance on land uplift isobases, the divisions made by Äyräpää were shown to be rather general in terms of geographic variation, as well as overlapping in terms of chronological sequence (Siiriäinen 1972, 15–16). Furthermore, subsequent archaeological shore displacement chronologies came to utilise the method known as radiocarbon dating, and, as result, the chronology gained greater precision (Meinander 1971; Siiriäinen 1973).

If evolution theory and geology provided the scientific background for the relative dating of archaeological finds, advances in particle physics provided a similarly revolutionary turning point for chronology in providing a method for ascertaining absolute rather than relative dates for archaeological finds. Radiocarbon dating was introduced in archaeology after its invention by Willard Libby in the late 1940s. The method is based on the assumption that the amount of ^{14}C isotopes occurring naturally in the earth's atmosphere is constant, and that the amount of ^{14}C present in all carbon based materials such as living organisms is proportionate to the amount of ^{14}C in the atmosphere. Because the half-life of ^{14}C is assumed to be 5660 ± 30 years, the age of the measured sample can be calculated by comparing the amount of remaining ^{14}C to the amount of ^{14}C in modern samples. Interestingly, the first tests on the new method were conducted on archaeological materials of known age, most importantly on Egyptian samples that could also be dated with the Pharaonic chronology (see Olsson 2009 for the history of radiocarbon dating).

One of the first Finnish archaeologists to utilise radiocarbon dating was professor Carl Fredrik Meinander. Meinander introduced the method to his Finnish colleagues in the early 1950s, and later published the first radiocarbon datings of Finnish archaeological material in 1971 (Meinander 1951, 1971). Meinander (1971) reported 18 datings from six sites mostly from southern Finland, including five samples from the Eura Honkilahti Kolmhaara dwelling site. One particular sample of charcoal from the Kolmhaara site was dated to 5440 ± 160 BP (Meinander 1971, 5). Because Typical Comb Ware was found in the same archaeological context, Meinander concluded that the dating for that type of pottery in that area would correspond with the dating of roughly 5500 BP, calculated from AD 1950, which is held as the value for present day in radiocarbon dating.

All in all, Meinander's suggested dates for Typical Comb Ware in Finland ranged between 4980 ± 150 BP and 5510 ± 170 BP (see also Siiriäinen 1972, 16; Siiriäinen

1973, 11). These dates were not only significantly earlier than the 4200–3950 BP originally estimated by Äyräpää, but they were also earlier than the 4250–3950 BP dating for Typical Comb Ware reported by Meinander himself only six years earlier (Meinander 1971, 7). The radiocarbon method seemed to provide datings that were much older than those derived through relative typological chronology. Furthermore, in the course of the 1970s, the radiocarbon method was further developed by introducing calibration of radiocarbon dates (Renfrew 1973; Taylor 1997). Because the amount of radioactive ^{14}C isotope in the earth's atmosphere is not constant, the age of the dated sample cannot be calculated from the most probable half age of ^{14}C. Dendrochronology, the dating of tree rings, was mostly used in order to provide samples with known calendrical ages for the calibration of radiocarbon dating results. In general, the calibration tended to make young dates even younger, while pushing the chronology even further back in the older end (see also Seger 1991).

Although the earliest calibration curves were already established by the 1970s, for a long time, the radiocarbon dates were given in archaeological literature in uncalibrated dates due to prejudices regarding the reliability and variability of the available calibration methods. The mid-1980s mark a turning point in this respect as a consensus was reached regarding the use of calibration methods available and a host of calibration curves for different periods were published in an issue of Radiocarbon. Interestingly, however, calibrated dates were not widely used in Finnish archaeological literature until the 1990s (Hiekkanen et al. 1988; Masonen et al. 1988; Seger 1991); see also Jungner (1995). In the late eighties, a calibration curve provided by Pearson et al. (1986) would have been used to date Typical Comb Ware. Consequentially, the 5440 BP dating from Kolmhaara would correspond to about 6290 cal BP, and would therefore be much older than the dates for earliest Typical Comb Ware. According to currently available calibrated dates, the beginning of ceramics use in south-western Finland dates to about 7100 BP, while Typical Comb Ware does not appear until some time around 6000 BP (Pesonen and Leskinen 2009; Pesonen et al. 2012; Herva et al. 2017, 26) and references therein. The age of the burnt tree, which could have easily been over 200 years old at the time of its felling, could explain this discrepancy.

Because the estimated actual dates for the relative chronologies of Finnish Stone Age cultures were constructed with reference to other European and Scandinavian archaeological cultures, the same ageing effect took place in their case as well. Many of the European prehistoric datings, which were based on typological artefact comparisons with the material culture of Mediterranean and Egyptian civilisations with known calendrical ages, proved to be older than previously estimated (Renfrew 1973). It is therefore understandable that radiocarbon dating, as well as evolution theory, and lately the study of ancient DNA, have been characterised as scientific revolutions, or definitive turning points in terms of methodological advances in archaeology (see also Kristiansen 2014). In this sense, the initial introduction of radiocarbon dating in archaeology can be characterised as the 'first radiocarbon revolution', while the calibration of those dates resulted in a second radiocarbon revolution (Renfrew 1973).

While the aforementioned dating methods have no doubt yielded results that can in many ways be considered more exact or less ambiguous than the comparative typological method, they nevertheless include several possible sources of error. In the case of radiocarbon dating, the aforementioned need for calibration is not the only possible source of misinformation. As became evident with the Kolmhaara case, the dated sample itself could be much older than the artefact found in the same context. One issue then regards possible contamination. The dated sample could include traces of older and/or younger materials than the actual age of the activity being dated. The context being dated could be mixed, and contain materials covering a long time period. When radiocarbon dates differ significantly from the expected results, they are often treated as contaminated and dismissed as anomalous outliers. Inversely, research can come to rely on dating results too much. When the initial radiocarbon dating results conducted on Egyptian materials came in, the Pharaonic chronology that was established through culture-historical methods was thought to be off. As it turned out, the dating method itself needed tweaking.

The establishment of archaeological chronologies with the help of land uplift introduces similar sources of error. While from very early on, it was evident that the speed of land uplift is gradually slowing down, it was also observed that the rate of uplift is not similar over the whole landmass. The effect was accounted for in archaeological research, and to this end Äyräpää for instance indicated the elevations for finds in percentages from the regional Litorina maximums reported by Ramsay. Similarly, it was also observed that significant regressions had taken place is certain areas. As consequence, large inland water bodies, such as the Ancylus lake, had formed and covered much of the region of present day Finland. Due to these regional irregularities in land uplift, many sites that were once on dry land have later submerged as result of various natural processes (Ramsay 1926; Saarnisto and Siiriäinen 1970; Koivisto 2017).

Much of the above mentioned sources of error can be gradually eliminated over time with the accumulation of data and the development of the adopted method. However, even the most precise methods introduce anomalies that cannot be explained with the processes that the method aims to model. The most common critique of the methods discussed in this chapter, namely evolution theory and land uplift models, has been that they emphasize large-scale natural processes over more specific and regional cultural processes and therefore tend to treat culture as a natural product and dismiss the human factor. It is then often noted that explanations based on natural processes fail in anticipating the somewhat less predictable effects of human activities in particular areas or archaeological sites. In the case of land uplift modelling, anomalies that the model itself is unable to explain or predict are at times introduced due to the effects of human activities such as land use intensification or damming. In addition to predicting and explaining land uplift one important application of land uplift models then is their ability to detect or discover anomalies that could have resulted from human activities rather than land uplift.

The effects of land use and modification on the prehistoric landscape were most likely minimal during the Stone Age. Stone Age subsistence in south-western Finland was mostly based on hunting, gathering, and fishing. As was pointed out above, the construction of Stone Age chronology has been founded on the idea that Stone Age populations were shorebound. This is most probable because the subsistence of mobile Stone Age populations was highly dependent on marine food sources, especially in the summer months. Lately, this view has become even better established as result of more careful collecting of small and easily perishable fish bone fragments on archaeological excavations in recent years. The earliest reliable signs of farming on a significant scale are from the early Bronze Age, and reflect the cultural influence of the Corded Ware Culture that introduced animal husbandry in southwestern Finland in the late Neolithic (Haggrén et al. 2015, 113). Because the Corded Ware culture as well as the Bronze Age economies relied more on animal husbandry and small-scale farming than hunting and gathering, the groups would be more sedentary. By the same token, Corded Ware and Bronze Age dwelling sites would more often be located further inland from coastal areas, and in land suitable for pasture and farming. The effects of mobile economy and intensifying land use resulted in clearing of forests and increasing erosion of the landscape. During the late Neolithic, however, the climate started to cool down and, as result, farming might have been significantly less intensive than in other parts of Scandinavia. As consequence of this cooling down event, farming in Finland intensified and became to form a significant part of subsistence in the Iron Age around AD 400–600, which, coupled with the increasing population, resulted in further clearing of the landscape, especially in the southern coastal areas (Haggrén et al. 2015, 131, 164). It has, however, been suggested that the earliest small-scale experiments of agriculture in Finland took place already during the early Neolithic, starting around 6000 BP, and therefore also corresponding with the Typical Comb Ware (Mökkönen 2010). Most likely these early signs of agriculture, identifiable only through cereal pollen data, reflect sporadic, small-scale, and experimental farming, and should therefore be seen as evidence of a long and gradual process in the course of which human economies became deeply entangled with the intensifying farming and clearing of the landscape toward the middle ages and the end of the prehistoric period.

References

Ailio J (1922) Fragen der russischen Steinzeit. K. F. Puromiehen Kirjapaino Oy

Aspelin JR (1875) Suomalais-ugrilaisen muinaistutkinnon alkeita, vol 51. Suomalaisen kirjallisu-uden seura

Aspelin JR (1885) Suomen asukkaat pakanuuden aikana. K. E. Holm, Helsinki

Baudou E (2005) Kossinna meets the Nordic archaeologists. Curr Swed Archaeol 13:121–140

Bulkin VA, Klejn LS, Lebedev GS (1982) Attainments and problems of Soviet archaeology. World Archaeol 13(3):272–295. https://doi.org/10.1080/00438243.1982.9979834

Clark J, Darlington J, Fairclough G, Mikkonen-Hirvonen S, Tiitinen T (2003) Polkuja euroop-palaiseen maisemaan: pathways to the Cultural Landscape hanke 2000–2003, finnish ed edn. PCL 2003

Edgren T (1966) Jäkärlä-gruppen. En västfinsk kulturgrupp under yngre stenålder. Suomen muinais-muistoyhdistyksen aikakauskirja 64, University of Helsinki

Edgren T (1984) On the economy and subsistence of the battle-axe culture in Finland. Fenno-ugri et slavi 4:9–15

Eronen M, Gunnar G, van de Plassche O, van der Plicht J, Rantala P (1995) Land uplift in the Olkiluoto-Pyhäjärvi area, Southwestern Finland, during the last 8000 years. Technical report, Nuclear Waste Commission of Finnish Power Companies, Eurajoki

Eronen M, Glückert G, Hatakka L, van de Plassche O, van der Plicht J, Rantala P (2001) Rates of Holocene isostatic uplift and relative sea-level lowering of the Baltic in SW Finland based on studies on isolation contacts. Boreas 30(1):17–30. https://doi.org/10.1111/j.1502-3885.2001.tb00985.x

Europaeus A (1923) Säkkijärven Ravin kivikautiset asuinpaikat. Suomen Museo 29:20–31

Europaeus A (1926) Stenålderskeramik från kustboplatser i Finland. Suomen Muinaismuistoyhdistyksen Aikakauskirja XXXVI(1):45–77

Europaeus-Äyräpää A (1930) Die relative Chronologie der steinzeitlichen Keramik in Finnland I-II. Acta Archaeol I(165–190):205–220

Gamble C, Kruszynski R (2009) John Evans, Joseph Prestwich and the stone that shattered the time barrier. Antiquity 83(320):461–475. https://doi.org/10.1017/S0003598X00098574

Gräslund B (1987) The birth of prehistoric chronology. Cambridge University Press

Haggrén G, Halinen P, Lavento M, Raninen S, Wessman A (2015) Muinaisuutemme jäljet. Suomen esi-ja varhaishistoria kivikaudelta keskiajalle. Gaudeamus, Helsinki

Herva VP, Mökkönen T, Nordqvist K (2017) A northern Neolithic? Clay work, cultivation and cultural transformations in the boreal zone of north-eastern Europe, c.53003000bc. Oxf J Archaeol 36(1):25–41. https://doi.org/10.1111/ojoa.12103

Hiekkanen M, Jungner H, Matiskainen H (1988) A sewn boat from Lake Mammosenjärvi in Puumala, Eastern Finland. Finsk Museum, pp 41–51

Hildebrand H (1873) Den Vetenskapliga Fornforskningen, hennes uppgift behof och, rätt. Tryckt hos A.L. Norman

Huurre M (1991) Satakunnan kivikausi,Satakunnan historia 1. Satakunnan Maakuntaliitto

Jungner H (1995) Dating in Archaeology. Fennosc Archaeol XII:37–38

Kivisalo N (2008) The Late Iron Age Bear-Tooth Pendants in Finland: Symbolic Mediators between Women, Bears, and Wilderness? Temenos 44(2):263–291. http://ojs.tsv.fi/index.php/temenos/article/download/4591/12446

Klindt-Jensen O (1975) A history of scandinavian archaeology. Thames and Hudson, London

Koivisto S (2017) Archaeology of Finnish Wetlands With Special Reference To Studies of Stone. Ph.D. dissertation, University of Helsinki

Kristiansen K (2014) Towards a new paradigm? Curr Swed Archaeol 22(Barrett):11–34

Lanting JN, Aerts-Bijma aT, van Der Plicht J, (2001) Dating of cremated bones. Radiocarbon 43(2A):249–254. https://doi.org/10.1017/S0033822200038078, http://www.rug.nl/research/portal/publications/dating-of-cremated-bones(3133775a-3e86-49c6-bef1-bc509f1c532b).html

Lehtosalo-Hilander PL (2000) Kalastajista kauppanaisiin: Euran esihistoria. Euran kunta, Eura

Marila M (2018) Finnish reactions to new archaeology. Fennosc Archaeol

Masonen J, Kankainen T, Zetterberg P (1988) Ancient land communications research in Finland. Fennosc Archaeol V:79–104

Meinander CF (1951) Om de förhistoriska dateringarnas vansklighet. Nordenskiöld-samfundets Tidskrift XI:3–14

Meinander CF (1971) Radiokarbondateringar till Finlands stenåldern. Föredrag hållet vid Finska Vetenskaps-societetens sammanträde den 20 April 1970. Societas scientiarum Fennica yearbook 48(B 5):1–14

Montelius O (1884) Den förhistoriske fornforskarens metod och material. Antiqvarisk tidskrift för Sverige 8(3):1–28

Montelius O (1899) Typologien eller utvecklingsläran tillämpad på det mänskliga arbetetNo Title. Svenska Fornminnesföreningens Tidskrift 10(3):237–268

Müller S (1884) Mindre bidrag til den forhistoriske archæologis methode. In: Aarbøger for Nordisk Oldkyndighed og Historie, pp 161–216

Müller S (1897) Vor Oldtid. Danmarks forhistoriske archæologi. Det Nordiske Forlag, Copenhagen

Mökkönen T (2010) Kivikautinen maanviljely Suomessa. Suomen Museo 2009:5–38

Nordqvist K, Mökkönen T (2015) Äyräpää's Typical Comb Ware: and umbrella term for the early 4th millennium BC pottery in Northeastern Europe? Fennosc Archaeol XXXII:151–158

Olsson IU (2009) Radiocarbon dating history: early days, questions, and problems met. Radiocarbon 51(1):1–43. https://doi.org/10.2458/rc.v51i1.3477

Pälsi S (1915) Riukjärven ja Piiskunsalmen kivikautiset asuinpaikat Kaukolassa. University of Helsinki

Pesonen P, Oinonen M, Carpelan C, Onkamo P (2012) Early subnealithic ceramic sequences in eastern Fennoscandia a Bayesian approach. Radiocarbon 54(3–4):661–676

Pearson GW, Pilcher JR, Baillie MGL, Corbett D, Qua F (1986) High-precision ^{14}C measurement of Irish oaks to show the natural ^{14}C variation from AD 1840 to 5210 BC. Radiocarbon 28(2B):911–934

Pesonen P, Leskinen S (2009) Pottery of the stone age hunter-gatherers in Finland. In: Jordan P, Zvelebil M (eds) Ceramics before farming: the dispersal of pottery among prehistoric Eurasian hunter-gatherers. Left Coast Press, CA, pp 299–318

Pukkila J (2011) Liikettä ajassa ja paikassa–Lounais-Suomen muinaisrannat tarkastelussa. In: Uotila Kari (ed) Avauksia Ala-Satakunnan esihistoriaan. Eura Print Oy, Eura, pp 111–132

Raike E (2012) Maiseman kruunut: hautaröykkiöt Satakunnassa. In: Satakunnan kulttuuriympäristöt eilen, tänään, huomenna, Satakunnan Museo, p 343

Räty J (1988) Arkeologisen kokoelman alku Satakunnan Museossa. Muinaistutkija 3

Ramsay W (1920) Litorinagränsen i sydliga Finland. Geologiska Föreningen i Stockholm Förhandlingar 42(5):243–263

Ramsay W (1926) Eustatic changes of level and the neolithicum. Suomen Muinaismuistoyhdistyksen aikakauskirja XXXVI(2):3–18

Renfrew C (1973) Before civilization: the radiocarbon revolution and prehistoric Europe. Knopf, NY

Saarnisto M, Siiriäinen A (1970) Laatokan transgressioraja. Suomen Museo pp 10–22

Salminen T (2003) Suomen tieteelliset voittomaat: Venäjä ja Siperia suomalaisessa arkeologiassa 1870–1935. Ph.D. thesis, University of Helsinki

Salo U (1981) Satakunnan historia I,2 : Satakunnan pronssikausi. Satakunnan Maakuntaliitto

Salo U (2000) Ihmisen jäljet Satakunnan maisemassa. Suomalaisen kirjallisuuden seura, Helsinki

Salminen T (2013) Arkeologia tieteenä: tutkijoiden keskustelua 1900-luvun alkukymmeninä. Muinaistutkija 2:25–44

Seger T (1991) Ten thousand years of Finnish prehistory. A tentative calibration of the earliest radiocarbon dates. Finskt Museum, pp 14–23

Siiriäinen A (1969) Über die Chronologie der steinzeitlichen Küstenwohnplatze Finnlands im Lichte der Uferverschiebung. Suomen Museo 76:40–73

Siiriäinen A (1972) A gradient/time curve for dating stone age shorelines in Finland. Suomen Museo 79:5–18

Siiriäinen A (1973) Studies relating to shore displacement and stone age chronology in Finland. Finskt Museum 80:5–22

Siiriäinen A (1989) Aarne Äyräpää ja kampakeramiikka. In: Huurre M (ed) Aarne Äyräpää. Tutkija, opettaja, kansalainen. 1980-luvun näkökulmia, National Board of Antiquities, Helsinki, pp 29–40

Soikkeli-Jalonen A (2016) Pronssikautiset pronssiesineet Suomessa. Pro gradu thesis, University of Turku

Tallgren AM (1934) Oman itsensä kanssa painiskeleva muinaistiede. Kalevalaseuran vuosikirja 14:200–211

Tallgren AM (1937) The method of prehistoric archaeology. Antiquity 11(42):152–161. https://doi.org/10.1017/S0003598X0001259X

Taylor R (1997) Radiocarbon dating. In: Taylor R, Aitken MJ (eds) Chronometric dating in archae-ology. Plenum Press, NY, pp 65–96
Tiitinen T (2011) Liikettä ajassa ja paikassa–Lounais-Suomen muinaisrannat tarkastelussa. In: Kari U (ed) Avauksia Ala-Satakunnan esihistoriaan. Eura Print Oy, Eura, pp 47–80

Chapter 3
Modelling of Postglacial Landscape Development

Abstract The postglacial land uplift process has been modelled using two different approaches: by modelling the geodynamics of the earth's crust (also referred to as Glacial Isostatic Adjustment (GIA) modelling) or by fitting mathematical models to existing archaeological and geological data (referred to as semi-empirical modelling). Although the semi-empirical models are not based on the physical properties of the earth's crust, they are easy to implement and can adapt better to local variations when compared to the GIA models. Semi-empirical models are fitted to the ice retreat data, eustatic sea level dynamics and lake isolation data on past shoreline displacement. As most of these data sources involve uncertainties, the land uplift process can be modelled probabilistically using the Monte Carlo method.

Keywords Land uplift modelling · Shoreline displacement · Ice retreat
Eustatic model · Optimization

Land uplift models can be either based on the data obtainable from past shoreline displacement (so-called semi-empirical models) or on the knowledge about the properties of the earth's crust and the geodynamical processes underlying land uplift (geodynamical models). Geodynamical modelling of land uplift, also referred to as Glacial Isostatic Adjustment (GIA) modelling, involves finding the balance between sea level and landmass. GIA models combine all the known forces affecting land uplift such as ice load and tidal forces, for example. The model seeks an equilibrium between all these forces taking into account the earth's physical properties. These equations could be solved either directly, requiring enormous computational power and computer memory capacity, or using finite element techniques. In these types of models, computation is performed over a one-, two- or three-dimensional mesh grid. Peltier et al. were the first to note the importance of global finite element models to solve the sea-level equations (Peltier et al. 1978). Variables describing the physical properties of the earth such as crust tensions and lithosphere temperature, for example, are estimated and input to the mesh grid model. The unknown model parameters are then estimated and the grid model is updated iteratively to find the equilibrium (Gerya 2009; Sabadini and Vermeersen 2004). However, the GIA models tend to be very complex so that either a limited area of the earth's crust can be described at a time or the grid is too coarse for the results to be applied locally. Also, the models usually make assumptions that might not be justified.

Development of GIA models is an iterative process of modifying the starting parameters as new information becomes available, re-modelling, and analyzing the modelling results. Many developers such as Richard Peltier, Kurt Lambeck, Jerry Mitrovica, Glenn Milne, Giorgio Spada, Patrick Wu and Georg Kaufmann, to name a few, have worked with their groups over several decades for refining their GIA models. In many cases, cooperation between the different research groups has led to improved models when compared to earlier model versions. These GIA models are usually global because the earth's crust must be considered as a whole and therefore, it means that glaciers other than those of Fennoscandia, as well as other global phenomena, must be taken into account. However, it is possible to compute the Fennoscandian area of the global GIA models with an isolated smaller version of the model with carefully selected boundary conditions. A good review of GIA models and a brief history of GIA modelling is presented in Whitehouse (2009).

Due to the restrictions mentioned above, the grid resolution of geodynamical models is usually of the order of several tens to hundreds of kilometres. This means that there may be one computational point in the whole Satakunta area. This poses a problem for historical event identification in the Lower Satakunta area. For that reason, a data-driven semi-empirical modelling approach might be a good alternative to find local features of the land uplift process.

Short-term linear extrapolations of the land uplift process can be made directly from the datasets obtained, for example, from precise GPS stations such as in Lidberg et al. (2010), Poutanen et al. (2010). In Vestøl (2006) precise levelling, tide-gauge and GPS station data from Fennoscandia were fitted to a model using a least squares solution. The time range of these datasets is yet too narrow to make long-term extrapolations. Also, miniscule local movements seen in the data cannot be extended to estimate all local small movements at the Fennoscandian scale. In Scherneck et al. (2001), Vestøl et al. (2016), Scandinavian BIFROST (Baseline Inferences for Fennoscandian Rebound Observations, Sea Level and Tectonics) land uplift time series containing the whole of Fennoscandia was obtained by using the GPS network for isostatic rebound information. In Kollo and Vermeer (2010), BIFROST data were used for modelling both horizontal and vertical movements related to the land uplift process to obtain lithospheric thickness estimates. Combined gravimetric (obtained from the GRACE satellite, for example) and GPS network data were used to estimate the current uplift in Fennoscandia in Müller et al. (2012), Timmen et al. (2004).

To develop long-term semi-empirical land uplift models, geological and archaeological data can be used. Geological data consist of lake isolation data, where the lake bottom sediment layers are examined to find the layer where the saltwater algae are replaced by freshwater algae. This layer is then radiocarbon dated. Also, the lowest organic layer of ponds and peat bogs can be dated to get estimation of bog or pond isolation (Eronen et al. 2001). In Ojala et al. (2013), a database of ancient shoreline information was presented for the area of Finland. The data were, e.g. used to construct shorelines for specified time points. Archaeological data comes from human remains and artefacts such as ceramics or fire-place charcoal, for example (Bågenholm 1995; Tallavaara et al. 2010). It can always be questioned if the remains are carried to the place of their finding afterwards, however, usually the fire-places and burial sites are reliable data sources when the oldest dated sample from a particular place is considered for land uplift modelling. Both fireplaces and burial sites have been built on dry land and this information can be used when designing semi-empirical land uplift models.

Combining lake isolation data reveals that the Holocene uplift rate is not linear (Eronen et al. 2001). Tore Påsse developed an arctangent based rebound model that is loosely based on eustatic changes, ice recession timing and earth's structure based rebound parameters (Påsse 2001). The parameters of this model are determined by data fitting and the underlying mathematical function can be adjusted to follow local data variations very smoothly. However, the model is only remotely related to the real geophysical properties of the earth's crust.

3.1 Semi-empirical Land Uplift Model by Tore Påsse

The land uplift model considered in this work is based on the shoreline displacement information and was originally proposed in Påsse (2001). The model has been updated with new data on ice retreat, eustatic sea level rise and the depth of the earth's crust. Also, new data on the location of the shoreline has been gathered from several sources with the addition of archaeological data of early human settlements in the coastal areas.

The shore-level displacement estimate S is a sum of two components, the total glacio-isostatic uplift U and the eustatic sea level rise E:

$$S = U - E \tag{3.1}$$

The glacio-isostatic uplift U can be further divided into two components: U_s (slow uplift) and U_f (fast uplift). The components of the model are presented in Fig. 3.1.

The slow uplift can be expressed as

$$U_s = \frac{2}{\pi} A_s \left(\arctan \frac{T_s}{B_s} - \arctan \frac{T_s - t}{B_s} \right), \tag{3.2}$$

where A_s is called the download factor (in metres), B_s is called the inertia factor (in years^{-1}), T_s is the time for the maximal uplift rate (in years) and t denotes the time variable (in years). The fast uplift can be expressed by

$$U_f = A_f e^{-0.5 \left(\frac{t - T_f}{B_f} \right)^2}, \tag{3.3}$$

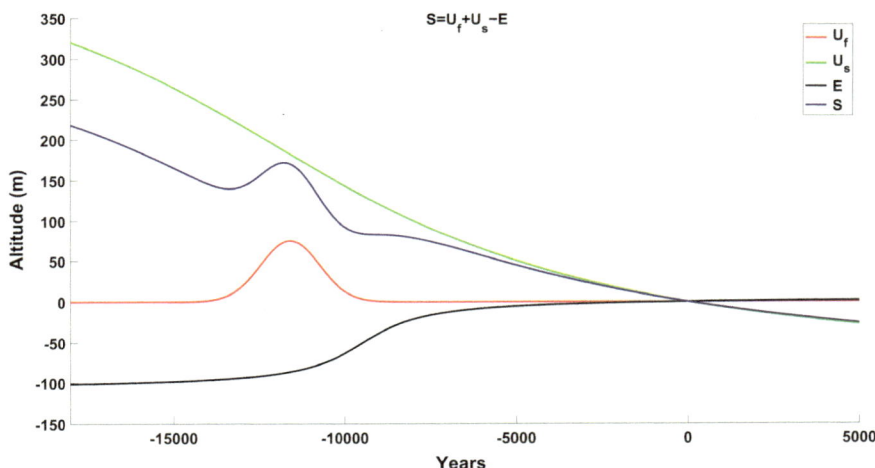

Fig. 3.1 An example of shore-level displacement S, land uplift U_s and U_f and eustatic sea level rise E based on the land uplift model by Påsse (2001)

where A_f represents total subsidence (in metres), B_f is the inertia factor (in years^{-1}), T_f is the time of the estimated turning point from subsidence to uplift (in years) and t is the time variable (in years). In Fig. 3.1, the altitudes correspond to the components of the model and this can be done by choosing a certain reference level. The figure follows the convention used in Påsse (2001) where the reference point is set to shore level in AD 1950, which is common also in carbon dating.

The land uplift modelling challenge is the optimization of the A_s and B_s parameters of the slow uplift U_s. The fast uplift U_f took place over 10 000 years ago, so its impact in the Satakunta region is negligible and therefore it has been omitted from the modelling process. The optimization process of the land uplift model involves two steps. In the first step, the components and parameters of the model are modified or constrained based on what is known about the respective processes based on the literature. This includes

- determining the value of the T_s parameter based on findings of the ice retreat process
- modifying the eustatic sea level rise curve according to the historical evidence on the Baltic Sea level changes
- constraining the value of the B_s parameter according to what is known about the physical properties of the earth's crust.

In the second step, the A_s and B_s parameter values are fine-tuned based on various data relevant to historical shoreline displacement.

3.2   Ice Retreat in Fennoscandia

The T_s parameter in Påsse's uplift model corresponds to the time of ice sheet recession. To obtain the value for the T_s parameter a reconstruction of the ice sheet extent
as a function of time is needed. The ice sheet history in Fennoscandia has been a
subject for many studies where specific land formations related to the ice retreat have
been dated. Such work includes (Björck 1995) where the southern Baltic Sea history
from 13 000 to 8000 BP is analyzed in addition to the ice retreat. The ice retreat
in Fennoscandia has also been studied in Lokrantz and Sohlenius (2006), Lunkka
and Erikkilä (2012) based on, e.g. scratches in bedrock, ridge formations and pollen
analysis. The ice retreat in Scandinavia in this work is based on the data of two
studies: (Hughes et al. 2016; Stroeven et al. 2016). In Hughes et al. (2016) a reconstruction of the Eurasian ice sheets is presented whereas in Stroeven et al. (2016)
the Fennoscandian ice sheet was reconstructed. Both reconstructions are based on
collected published numerical dates constraining the timing of ice sheet advance and
retreat, and additionally geomorphological and geological evidence contained within
the existing literature. In Fig. 3.2 the model by Hughes et al. (2016) is presented.

3.3   Eustatic Model

The eustatic model was presented in Påsse (2001) as

$$E = \frac{2}{\pi} 56 \left(\arctan \frac{9500}{1350} - \arctan \frac{9500 - t}{1350} \right). \tag{3.4}$$

This model is based on an iterative process, where the difference between hypo-
thetical uplift curves and empirical shore-level curves was calculated. However, in
our work an alternative eustatic model based on water level data from several sources
is used. The main component of this model for the last 10,000 years is the eustatic
curve presented in Punning (1987) showing the changes in water level in the area
of the Baltic Sea. Other components of the alternative eustatic model include the
radiocarbon-dated coral data collected by Fairbanks (1989), Chappell and Polach
(1991), Bard et al. (1996) and the data from the past lake phases in the Baltic Sea
area by Björck (2008). Two eustatic curves, the one presenting the model by Påsse
(2001) and the another obtained by combining the coral data and taking into account
the effects of the lake phases, are presented in Fig. 3.3. A polynomial function was
used together with the segments of the lake phases (Björck 2008) and the water level
changes (Punning 1987) to create the alternative eustatic model. The model shows
notable rises in water level in the Baltic Sea area during the lake phases.

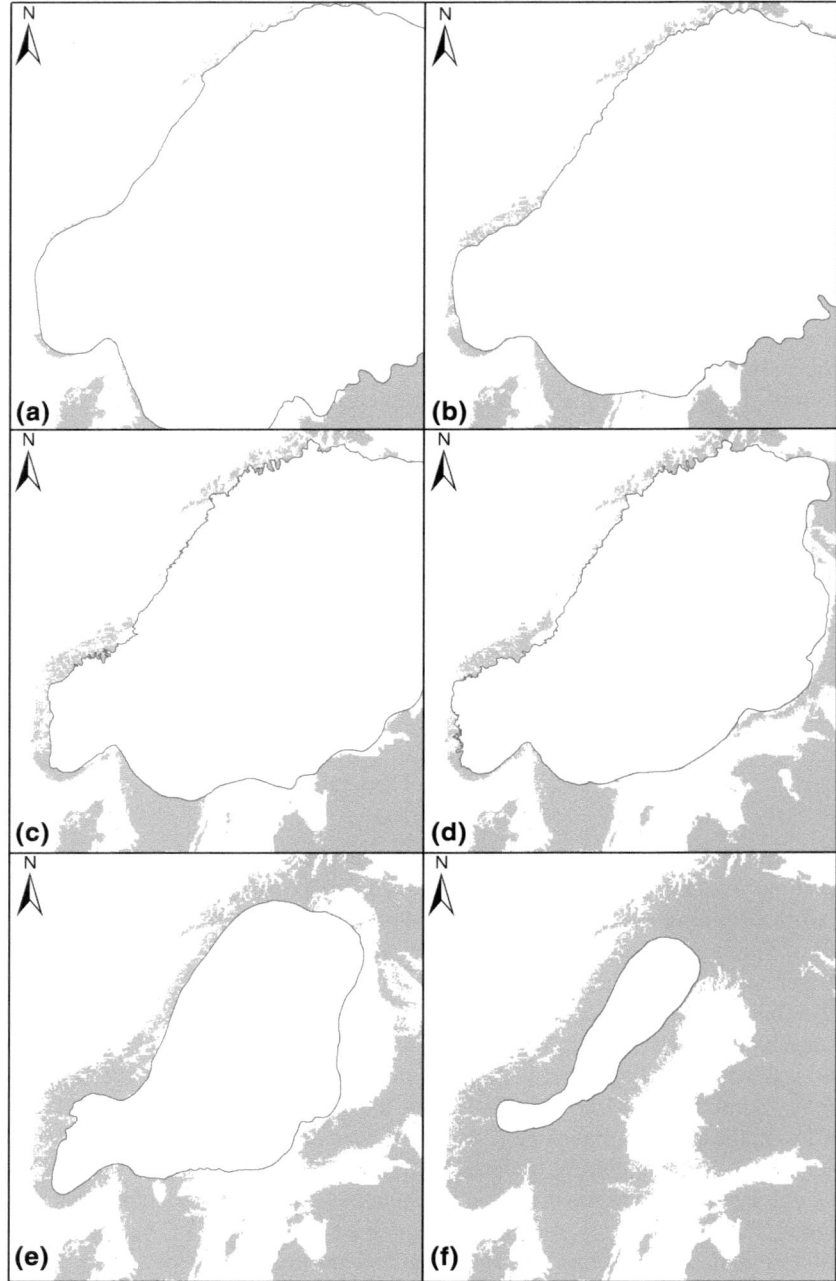

Fig. 3.2 Ice retreat model in Fennoscandia in **a** 15 000 BP, **b** 14 000 BP, **c** 13 000 BP, **d** 12 000 BP, **e** 11 000 BP and **f** 10 000 BP by Hughes et al. (2016). *Background map: Modified from the EU-DEM v1.1 by the European Environment Agency*

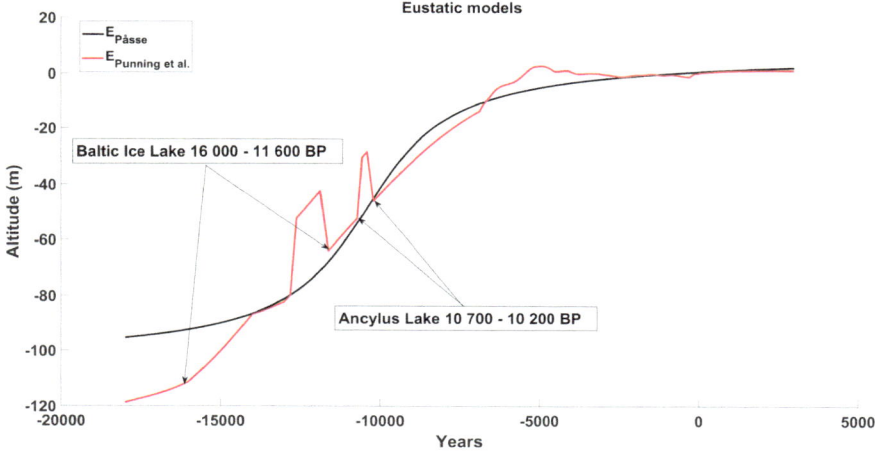

Fig. 3.3 The eustatic model by Påsse (2001) and the alternative model used in this work

3.4 Crustal Thickness Estimation

The inertia factor B_s in Eq. 3.2 can be expressed as Påsse (2001):

$$B_s = 302e^{0.067ct}. \tag{3.5}$$

In Eq. 3.5 ct is the crustal thickness of the earth. Crustal thickness is presented in the form of a Moho map (Mohorovičić discontinuity) describing the depth of the boundary between the earth's crust and the mantle. Påsse used the Moho map by Kinck et al. (1993) which shows similarity to a more detailed Moho map presented in Grad and Tiira (2009). The map by Grad and Tiira (2009) was used in this work.

3.5 Land Uplift Model Optimization

3.5.1 Source Data

The source data for the land uplift model parameter optimization includes two types of data: one collected from the bottom sediments of lake and mire basins, indicating the time when the environment changes from brackish water to fresh water, corresponding to lake isolation from the sea, and the other collected from, e.g. prehistoric dwelling sites, indicating the time when the particular location represented dry land.

The digital elevation model used in the study is the Elevation model 10 m by the National Land Survey of Finland. The grid size of the model is 10 m × 10 m and the model data is estimated to have an accuracy of 1.4 m on average (95%

confidence interval). The depth data for the lakes is based on the Lake and River Depth Profiles dataset by the Finnish Environment Institute. The data is in the form of depth contours, which were combined with the elevation model so that a uniform 10 m × 10 m grid was obtained.

In total the data set consists of 2406 radiocarbon dated points of which 1125 belong to the lake and mire isolation data set and 1281 to the archaeological data set (see Fig. 3.4). The majority of the points are located in Finland and Sweden, and thus they are in the most relevant area considering postglacial land uplift. A collection of lake and mire isolation data from Finland and Sweden have been presented in Vuorela et al. (2009). The majority of the data in Finland have been collected by Matti Eronen and Gunnar Glückert, some of this data is, for example, presented in Eronen et al. (2001). The age of Finnish peatlands has been studied in Mäkilä et al. (2013). In addition, data has been gathered from several sources, for example, from Estonia, Norway and Russia. The radiocarbon data of University of Helsinki has been collected in the ^{14}CARHU database (Junno et al. 2015). The database includes mainly archaeological data. Also some data were gathered from the collections of the Finnish Heritage Agency and the Swedish National Heritage Board that have put together databases of archaeological data in their respective countries.

3.5.2 Parameter Optimization

There are two major factors causing uncertainty in the land uplift parameter optimization process: the ^{14}C radiocarbon dating and the elevation value of the data points. Both the lake and mire isolation data set and the archaeological data set involves the use of ^{14}C radiocarbon dating procedure, from which a probability density function can be derived for the calendar age of the sample (Bronk Ramsey 2009). The uncertainty for the elevation value of the data point was taken care of by applying a Gaussian distribution to each elevation datum. A standard deviation of 3 m was used for covering this uncertainty due to, e.g. erosion. In the parameter optimization process, the archaeological data determines the upper limit for the water level.

Monte Carlo simulation with 1000 iterations was used to obtain the estimates of the U_s parameter values (Eq. 3.2). This kind of Monte Carlo based parameter optimization of a model can be considered as a curve fitting process where the minima of squared error of a set of random variables is sought. The random variables in the land uplift optimization process are the lake and mire basin isolation points and the archaeological data points including confidence limits. The new values are then taken randomly within their respective confidence limits and the most probable value for the model parameters are the maxima of the histograms. The land uplift model parameter optimization process is described in Fig. 3.5.

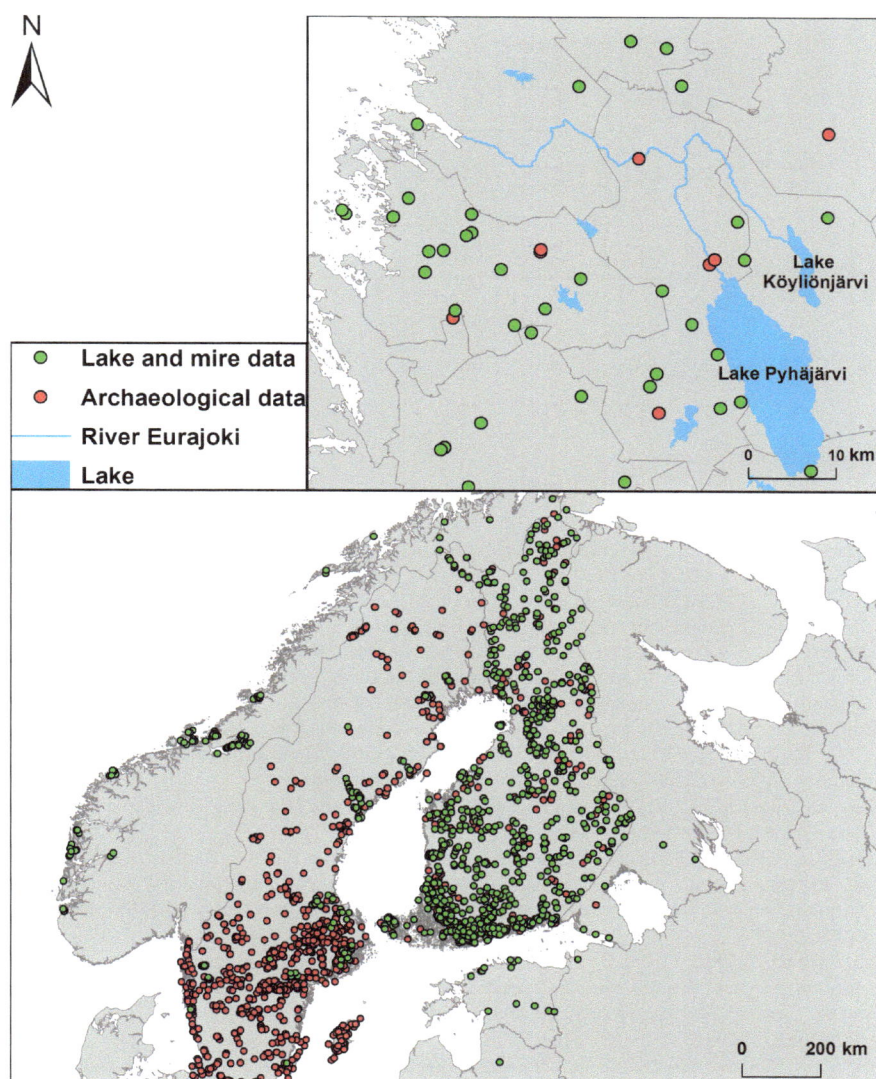

Fig. 3.4 Source data points used in the land uplift model optimization. *Background map in upper subfigure: National Land Survey of Finland. Lake and river network data in upper subfigure: Finnish Environment Institute. Background map in lower subfigure: Esri*

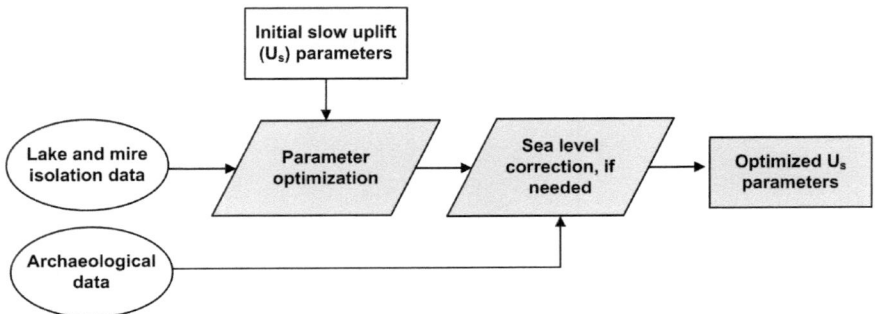

Fig. 3.5 The land uplift model parameter optimization process (modified from Pohjola et al. 2014)

References

Bard E, Hamelin B, Arnold M, Montaggioni L, Cabioch G, Faure G, Rougerie F (1996) Deglacial sea-level record from Tahiti corals and the timing of global meltwater discharge. Nature 382:241–244

Bågenholm G (1995) Corded ware ceramics in Finland and Sweden. Fennosc archaeol 12:19–23

Björck S (1995) A review of the history of the Baltic Sea, 13.0–8.0 ka BP. Quat Int 27(94):19–40. http://www.sciencedirect.com/science/article/pii/104061829400057C

Björck S (2008) The late quaternary development of the Baltic Sea Basin. In: von Storch H (ed) In assessment of climate change for the Baltic Sea Basin, chap A.2. Springer, Berlin, pp 398–407

Bronk Ramsey C (2009) Bayesian analysis of radiocarbon dates. Radiocarbon 51(01):337–360. https://doi.org/10.1017/S0033822200033865, https://www.cambridge.org/core/product/identifier/S0033822200033865/type/journal_article,1306.2418

Chappell J, Polach H (1991) Post-glacial sea-level rise from a coral record at Huon Peninsula, Papua New Guinea. Nature 349:147–149

Eronen M, Glückert G, Hatakka L, van de Plassche O, van der Plicht J, Rantala P (2001) Rates of Holocene isostatic uplift and relative sea-level lowering of the Baltic in SW Finland based on studies on isolation contacts. Boreas 30(1):17–30. https://doi.org/10.1111/j.1502-3885.2001.tb00985.x

Fairbanks RG (1989) A 17,000-year glacio-eustatic sea level record: influence of glacial melting rates on the younger dryas event and deep-ocean circulation. Nature 342:637–642

Grad M, Tiira T (2009) The Moho depth map of the European Plate. Geophys J Int 176(1):279–292. https://doi.org/10.1111/j.1365-246X.2008.03919.x, http://gji.oxfordjournals.org/cgi/

Gerya T (2009) Introduction to numerical geodynamic modelling. Cambridge University Press, New York. https://doi.org/10.1017/CBO9780511809101, http://ebooks.cambridge.org/ref/id/CBO9780511809101, arXiv:1011.1669v3

Hughes AL, Gyllencreutz R, Lohne ØS, Mangerud J, Svendsen JI (2016) The last Eurasian ice sheets-a chronological database and time-slice reconstruction, DATED-1. Boreas 45(1):1–45. https://doi.org/10.1111/bor.12142

Junno A, Uusitalo J, Oinonen M (2015) Radiocarbon dates of Helsinki University. www.oasisnorth.org/carhu

Kinck JJ, Husebye ES, Larsson FR (1993) The Moho depth distribution in Fennoscandia and the regional tectonic evolution from Archean to Permian times. Precambr Res 64:23–51

Kollo K, Vermeer M (2010) Lithospheric thickness recovery from horizontal and vertical land uplift rates. J Geodyn 50(1):32–37. https://doi.org/10.1016/j.jog.2009.11.006

Lidberg M, Johansson JM, Scherneck HG, Milne GA (2010) Recent results based on continuous GPS observations of the GIA process in Fennoscandia from BIFROST. J Geodyn 50(1):8–18. https://doi.org/10.1016/j.jog.2009.11.010, http://www.sciencedirect.com/science/article/pii/S0264370709001665

Lokrantz H, Sohlenius G (2006) Ice marginal fluctuations during the Weichselian glaciation in Fennoscandia, a literature review. Technical report, Swedish Nuclear and Waste Management Co, Stockholm. http://www.skb.se/upload/publications/pdf/TR-06-36.pdf

Lunkka JP, Erikkilä A (2012) Behaviour of the lake district ice lobe of the Scandinavian ice sheet during the younger dryas chronozone (ca. 12800–11500 years ago). Technical report April, Posiva Oy, Eurajoki, http://www.posiva.fi/files/2827/WR_2012-17web.pdf

Mäkilä M, Säävuori H, Kuznetsov O, Grundström A (2013) Age and dynamics of peatlands in Finland. Technical report, Geological Survey of Finland, Espoo, Finland

Müller J, Naeimi M, Gitlein O, Timmen L, Denker H (2012) A land uplift model in Fennoscandia combining GRACE and absolute gravimetry data. Phys. Chem. Earth 53–54:54–60. https://doi.org/10.1016/j.pce.2010.12.006

Ojala AE, Palmu JP, Åberg A, Åberg S, Virkki H (2013) Development of an ancient shoreline database to reconstruct the Litorina Sea maximum extension and the highest shoreline of the Baltic Sea basin in Finland. Bull Geol Soc Finland 85(PART 2): 127–144, https://doi.org/10.17741/bgsf/85.2.002

Påsse T (2001) An empirical model of glacio-isostatic movements and shore-level displacement in Fennoscandia. Technical report, Swedish Nuclear Fuel and Waste Management Co. http://www.skb.se/upload/publications/pdf/R-01-41.pdf

Peltier WR, Farrell WE, Clark JA (1978) Glacial isostasy element model. Tectonophysics 50:81–110

Punning YM (1987) Holocene eustatic oscillations of the Baltic Sea level. J Coastal Res 3(4):505–513

Pohjola J, Turunen J, Lipping T, Ikonen ATK (2014) Landscape development modeling based on statistical framework. Comput Geosci 62. https://doi.org/10.1016/j.cageo.2013.09.013

Poutanen M, Dransch D, Ivins ER, Klemann V, Kozlovskaya E, Kukkonen I, Lunkka JP, Milne G, Pascal C, Steffen H, Vermeersen B, Wolf D (2010) DynaQlim upper mantle dynamics and quaternary climate in cratonic areas. In: Cloetingh S, Negendank J (eds) New frontiers in integrated solid earth sciences. Springer Netherlands, Dordrecht, pp 349–372. https://doi.org/10.1007/978-90-481-2737-5, http://link.springer.com

Sabadini R, Vermeersen B (2004) Global dynamics of the earth. Kluwer Academic Publishers, Dordrecht. https://doi.org/10.1007/978-94-009-4800-6_8, http://link.springer.com/chapter/

Scherneck HG, Johansson JM, Vermeer M, Davis JL, Milne GA, Mitrovica JX (2001) BIFROST project: 3-D crustal deformation rates derived from GPS confirm postglacial rebound in Fennoscandia. Earth Planets Space 53:703–708. https://doi.org/10.1186/BF03352398

Stroeven AP, Hättestrand C, Kleman J, Heyman J, Fabel D, Fredin O, Goodfellow BW, Harbor JM, Jansen JD, Olsen L, Caffee MW, Fink D, Lundqvist J, Rosqvist GC, Strömberg B, Jansson KN (2016) Deglaciation of Fennoscandia. Quat Sci Rev 147:91–121. https://doi.org/10.1016/j.quascirev.2015.09.016

Tallavaara M, Pesonen P, Oinonen M (2010) Prehistoric population history in eastern Fennoscandia. J Archaeol Sci 37(2):251–260. https://doi.org/10.1016/j.jas.2009.09.035

Timmen L, Gitlein O, Denker H, Bilker M, Wilmes H, Falk R, Reinhold A, Hoppe W, Pettersen BR, Engen B, Engfeldt A, Strykowski G, Forsberg R, Observatory OS, Survey N (2004) Observing Fennoscandian geoid change for GRACE validation. In: Joint CHAMP/GRACE science meeting, Potsdam, 6 July–8 July 2004, pp 1–10

Vestøl O (2006) Determination of postglacial land uplift in Fennoscandia from leveling, tide-gauges and continuous GPS stations. J Geodesy 80(5):248–258. https://doi.org/10.1007/s00190-006-0063-7

Vestøl O, Ågren J, Steffen H, Kierulf H, Lidberg M, Oja T, Rüdja A, Kall T, Saaranen V, Engsager K, Jepsen C, Liepins I, Parseliunas E, Tarasov L (2016) NKG2016LU, an improved postglacial land uplift model over the Nordic-Baltic region. In: NKG working group of geoid and height systems

Vuorela A, Penttinen T, Lahdenperä AM (2009) Review of Bothnian Sea shore-level displacement data and use of a GIS tool to estimate isostatic uplift review of Bothnian Sea shore-level displacement data and use of a GIS tool to estimate isostatic uplift. Technical report, Posiva Oy, Eurajoki. http://www.posiva.fi/files/955/WR_2009-17web.pdf

Whitehouse P (2009) Glacial isostatic adjustment and sea-level change: state of the art report. Technical report TR-09-11, Swedish Nuclear Fuel and Waste Management Co, Stockholm. http://www.skb.se/upload/publications/pdf/TR-09-11.pdf

Chapter 4
Landscape Reconstruction in Lower Satakunta

Abstract The landscape reconstruction at four historical sites in Eura: Kolmhaara, Tyttöpuisto, Kuninkaanhauta and Luistari is presented in this chapter. The reconstruction focuses on the time periods of the archaeological findings at respective sites: Stone Age in the case of Kolmhaara and Tyttöpuisto, Bronze Age in the case of the Kuninkaanhauta and Iron Age in the case of Luistari. Landscape reconstruction results are discussed in relation with the radiocarbon datings of the findings at these sites. It is found that while for the Kuninkaanhauta the land uplift model locates the settlement at a favorable location with respect to the shoreline at approximately the same time period obtained using radiocarbon dating, there is a significant discrepancy between the two methods of estimating the age of the settlements for the Stone Age sites of Kolmhaara and Tyttöpuisto.

Keywords Landscape reconstruction · Radiocarbon dating · Land uplift

In this chapter, the land uplift model presented in Chap. 3 is used to reconstruct the historical shoreline, elevation profile and hydrological features at the vicinity of the four archaeological sites: Kolmhaara, Tyttöpuisto, Kuninkaanhauta and Luistari. In Finland, in the area of land uplift, it has been common to estimate the age of archaeological sites based on shoreline displacement assuming that people settled close to the shore. The validity of this assumption as well as the discrepancies between the uplift model and the datings of the archaeological findings in Western Finland have been thoroughly discussed in Tiitinen (2011). Tiitinen relied in her work on the reconstruction of elevation contours produced by Eronen et al. (2001). Since then, both the elevation model and the land uplift process have been refined considerably. Digital elevation models obtained from LiDAR data and hydrological modelling using GIS technology enable to visualize the landscape as a continuous surface instead of reconstructing individual elevation contours. Therefore, the sources of error such as the coarseness of elevation contours both in time and space, mentioned by Tiitinen as possible causes of discrepancies in dating, have a negligible effect in our reconstructions.

In Pohjola et al. (2014), we compared three land uplift models, (Påsse 2001; Vuorela et al. 2009; Punning 1987), all based originally on Påsse's mathematical model but using different data, procedures and eustatic models for parameter

J. Pohjola et al., *Historical Perspectives to Postglacial Uplift*,
SpringerBriefs in Geography, https://doi.org/10.1007/978-3-030-00970-0_4

49

optimization. The comparison was performed from the point of view of the future shoreline displacement as the aim was to model the landscape development in the vicinity of the future spent nuclear fuel repository in Olkiluoto. The main advancements of our land uplift model compared to those by Påsse and Vuorela are the usage of a more precise eustatic model by Punning and complementing the underlying database of archaeological findings. From Fig. 3.3 it can be seen, for example that the two eustatic models have several meters' discrepancy at about 5000 BP. In the following, visualizations of the landscape around the four sites are presented and dating of the sites is revisited based on the new land uplift model.

4.1   The Stone Age: Kolmhaara and Tyttöpuisto

The development of the Kolmhaara site from 5900 BP to 4700 BP is presented in Fig. 4.1. The oldest ^{14}C datings from the site are listed in Table 4.1. The reconstruction indicates that radical changes occurred in the landscape near Kolmhaara between 5500 and 5100 BP, when large land areas got lifted up from the sea. 5500 BP can probably be considered the earliest time the Stone Age sites shown in the figure became inhabitable. At that time, most of the sites would appear on islands. On the other hand, the landscape of the area in 5100 BP situates most of the sites on the coast of the mainland with Kolmhaara a couple of kilometers inland. As Kolmhaara is the largest of these sites, a logical explanation could be that it was inhabited first and when the shoreline shifted westwards, other sites were built on the coast with Kolmhaara remaining the 'base site'.

According to Table 4.1, the calibrated radiocarbon method dates the Kolmhaara site at 6230 ± 330 BP. In Tiitinen (2011) it was found that, by interpreting the elevation contours produced by Eronen et al. (2001), the site could be inhabited at the earliest in 5850 BP while our modified land uplift model makes the discrepancy even more significant. Therefore, our digital elevation model based landscape reconstruction does not explain away the anomaly but, instead, makes it more severe and further research has to address the question if this discrepancy is due to errors in radiocarbon dating or there are some unknown features in the land uplift at that period. Also, the eustatic model should still be revisited.

The development of Tyttöpuisto site from 5100 BP to 3900 BP is presented in Fig. 4.2. The oldest ^{14}C datings from the site are listed in Table 4.2. As for Kolmhaara, calibrated radiocarbon dating indicates the Tyttöpuisto site to be older, dating at 5800 ± 200 BP. According to the shore-level displacement model, the site reaches

Table 4.1 ^{14}C datings from the Kolmhaara site (Junno et al. 2015)	^{14}C BP	Cal BP (95.4% probability)
	5440 ± 160	6230 ± 330
	5420 ± 120	6180 ± 250

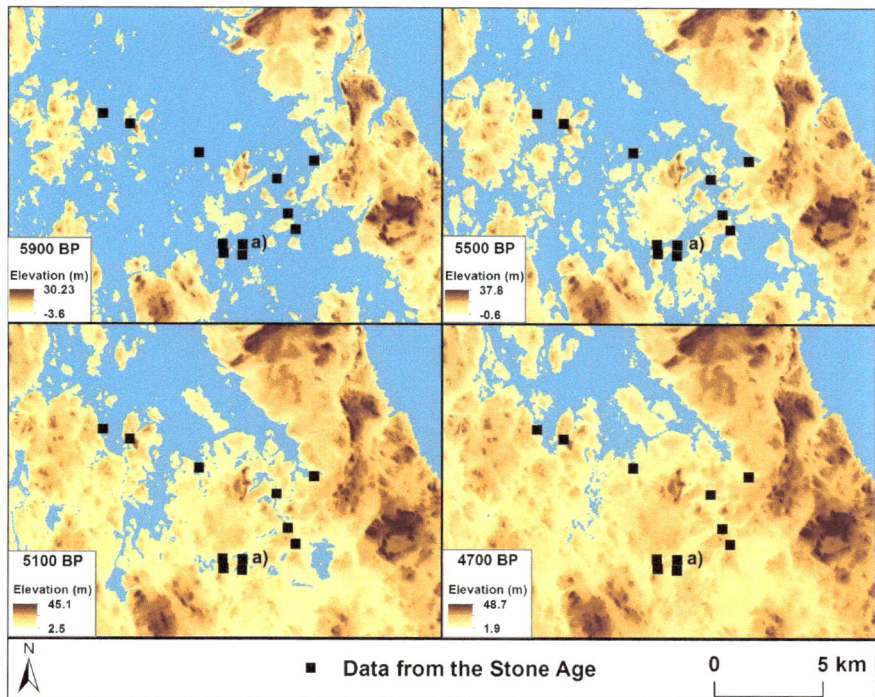

Fig. 4.1 Map of archaeological findings from the Stone Age and the shoreline displacement from 5900 to 4700 BP in Honkilahti, Eura. The location of the Kolmhaara Stone Age artefact is marked with (**a**). *Data points: Finnish Heritage Agency*

Table 4.2 ^{14}C datings from the Tyttöpuisto site (Junno et al. 2015)	[14]C BP	Cal BP (95.4% probability)
	5080 ± 100	5800 ± 200
	5070 ± 100	5790 ± 200

the shoreline at the earliest during the time period 5400–5300 BP. Based on our landscape reconstruction and assuming that the majority of the sites in the vicinity of Tyttöpuisto were originally located on the coast, the most probable time of their establishment is between 4700 BP and 4300 BP. In 4700 BP, most of the sites would be just on the coast with one of them on a small island. Taking into account that the land area is relatively flat and low and that annual fluctuations in sea level would exceed 2 m, 4300 BP would be more probable time for the establishment of the settlements.

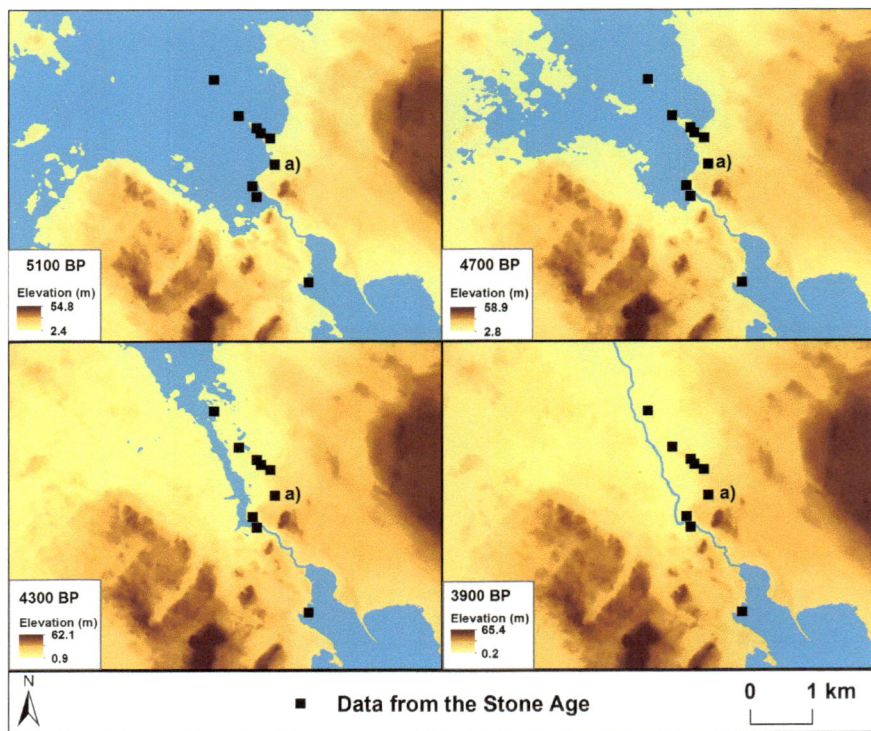

Fig. 4.2 Map of archaeological findings from the Stone Age and the shoreline displacement from 5100 BP to 3900 BP in Eura. The location of the Tyttöpuisto Stone Age Artefact is marked with (**a**). *Data points: Finnish Heritage Agency*

Table 4.3 ^{14}C dating in the vicinity of Kuninkaanhauta (Junno et al. 2015)	[14]C BP	Cal BP (95.4% probability)
	2470 ± 110	2550 ± 230

4.2   The Bronze and Iron Ages: Kuninkaanhauta and Luistari

The development of the Kuninkaanhauta site from 3000 BP to 2500 BP is presented in Fig. 4.3. A ^{14}C dating in the vicinity of the site is listed in Table 4.3. When looking at the landscape reconstruction, one can hardly imagine a better location for settlements than that of Kuninkaanhauta at about 3000 BP (or maybe slightly later). Most of the sites are scattered along the shore of a well-protected bay. From inland, the settlements would be surrounded by relatively steep hillsides. In the northern area, some of the settlements are located at the top of a hill with a good view to the sea.

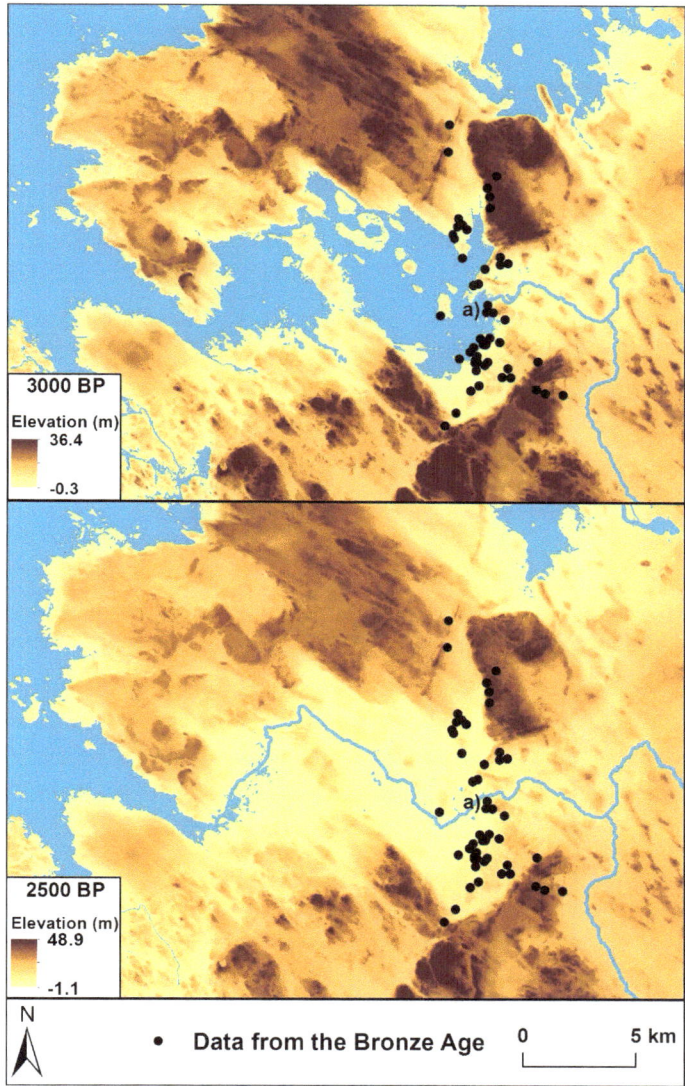

Fig. 4.3   Map of archaeological findings from the Bronze Age and the shoreline displacement from 3000 BP to 2500 BP in Panelia, Eura. The location of Kuninkaanhauta is marked with (**a**). *Data points: Finnish Heritage Agency*

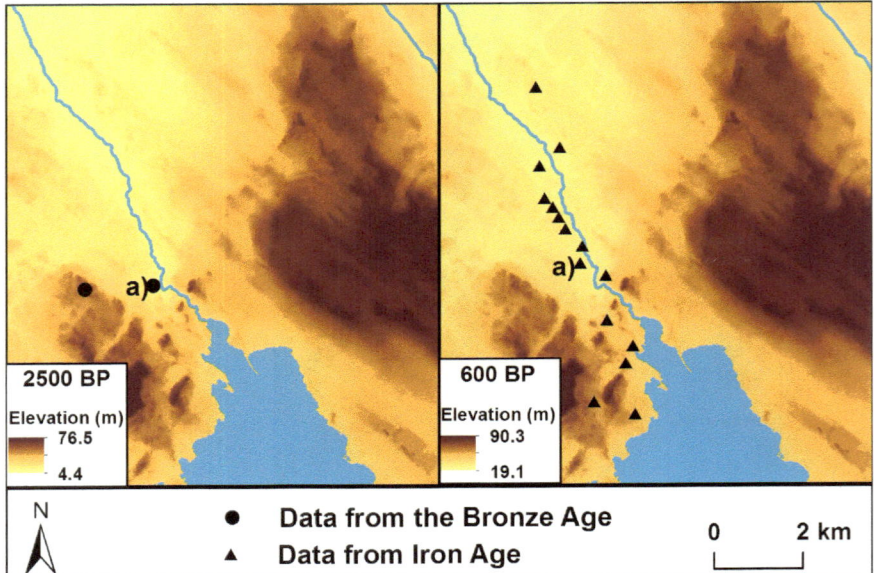

Fig. 4.4 Map of archaeological findings from the Bronze and the Iron Ages and the landscape in 2500 BP and in 600 BP in Eura. The location of the Luistari site is marked with (**a**). *Data points: Finnish Heritage Agency*

The calibrated radiocarbon method dates a dwelling site in the vicinity of Kuninkaanhauta to 2550 BP, which can be considered quite close to the about 2900 BP 'ideal-looking' landscape for the settlements. The slight discrepancy, in this case, is in the opposite direction compared to the Stone Age cases, Kolmhaara and Tyttöpuisto. It should be noted, however, that the oldest dated object does not necessarily come from the time the site was first inhabited.

The landscape of the Luistari site in 2500 BP (Bronze Age) and 600 BP (Iron Age) is presented in Fig. 4.4. The ^{14}C datings from the site are listed in Table 4.4. The Lake Pyhäjärvi has isolated from the sea about 5600 years ago, and the Eurajoki River has approximately maintained its location since then. Therefore, land uplift and resulting development of hydrological features of the landscape are not of much use when estimating the age of these historic sites. The reconstruction just reassures that the historic sites were, indeed, located along the Eurajoki River. We can also conclude that probably during the Bronze and Iron Ages the vicinity of sea was not of primary importance as far as there was a convenient access to the sea provided by a river. Also, Lake Pyhäjärvi provided food by fishing.

Table 4.4 ^{14}C datings from the Luistari site (Junno et al. 2015)

^{14}C BP	Cal BP (95.4% probability)
Bronze age	
2800 ± 80	2950 ± 190
2740 ± 90	2900 ± 180
Iron age	
640 ± 120	630 ± 170

References

Junno A, Uusitalo J, Oinonen M (2015) Radiocarbon dates of Helsinki University. www.oasisnorth. org/carhu

Eronen M, Glückert G, Hatakka L, van de Plassche O, van der Plicht J, Rantala P (2001) Rates of Holocene isostatic uplift and relative sea-level lowering of the Baltic in SW Finland based on studies on isolation contacts. Boreas 30(1):17–30. https://doi.org/10.1111/j.1502-3885.2001. tb00985.x

Påsse T (2001) An empirical model of glacio-isostatic movements and shore-level displacement in Fennoscandia. Technical report, Swedish Nuclear Fuel and Waste Management Co. http://www. skb.se/upload/publications/pdf/R-01-41.pdf

Pohjola J, Turunen J, Lipping T, Ikonen ATK (2014) Landscape development modeling based on statistical framework. Comput Geosci 62. https://doi.org/10.1016/j.cageo.2013.09.013

Tiitinen T (2011) Liikettä ajassa ja paikassa—Lounais-Suomen muinaisrannat tarkastelussa. In: Kari Uotila (ed) Avauksia Ala-Satakunnan esihistoriaan. Eura Print Oy, Eura, pp 47–80

Vuorela A, Penttinen T, Lahdenperä AM (2009) Review of Bothnian Sea shore-level displacement data and use of a GIS tool to estimate isostatic uplift review of Bothnian Sea shore-level displacement data and use of a GIS tool to estimate isostatic uplift. Technical report, Posiva Oy, Eurajoki. http://www.posiva.fi/files/955/WR_2009-17web.pdf

Punning YM (1987) Holocene eustatic oscillations of the Baltic Sea level. J Coastal Res 3(4):505–513

Chapter 5
Conclusions

Abstract The challenges of land uplift modelling and determining the age of pre-historic settlements are discussed in this chapter. Land uplift modelling is a valuable tool for re-iterating the datings of archaeological findings, but the model and its source data must be constantly reviewed and adjusted when new data becomes available. Also, the erosion and sedimentation processes and local variations in land uplift should be taken into account when refining the land uplift model. On the other hand, the results obtained by radiocarbon dating are challenged by phenomena such as the water reservoir effect, for example.

Keywords Radiocarbon dating · Marine reservoir effect · Erosion · Uncertainty

Seas, lakes, ponds and rivers have played an essential role to humans in history. They offered food by fishing as the inhabitants of the Stone Age settlements in Finland most probably were not involved in raising crops or domestic animals. Water bodies were also important for transportation. Therefore, it is no surprise that most of the old cities, centers, villages are founded near water. This is also true from the archaeological point of view. Figure 4.3 is a good example of the relation between historic human settlements and water.

The Kolmhaara and Tyttöpuisto sites studied in this work indicate that the calibrated radiocarbon datings tend to produce estimates 800–1200 years older than our landscape reconstruction results for Stone Age settlements. One possible reason for this is presented in Sect. 2.4 as the dated sample might be older than the artefact found from the surrounding area. Another reason for this discrepancy could be the 'Marine/Freshwater Reservoir Effect'. For example, in Dettman et al. (2015), Philippsen (2013), Reimer and Reimer (2006), it is observed that the samples that are below water appear to be approximately 200–600 years older depending on conditions than similar samples that are in dry land in terms of radiocarbon dating. That might be one reason why the dated burials in Kolmhaara seem to be in the Baltic Sea (see Fig. 4.1) when compared to the sea level calculated with the land uplift model. The burials in Kolmhaara are located in the bottom of a sandy ridge formation and water may have penetrated to the burials by means of the capillary adsorption mechanism of sand.

J. Pohjola et al., *Historical Perspectives to Postglacial Uplift*,
SpringerBriefs in Geography, https://doi.org/10.1007/978-3-030-00970-0_5

Modelling of shoreline displacement in the areas of post-glacial land uplift in the time scope of thousands of years is a good example of how mathematical modelling and archaeology can support each other. Estimating the age of archaeological arte-facts by means of the elevation of the particular sites has been common in studies focusing on the prehistory of Western Finland (Tiitinen 2011). The basic principle has been: the higher the place the older it must be. It has been realized, however, that this simple rule of thumb assumes that the sites were originally established at the coast. Recent advancements in GIS technology, remote sensing in acquiring digital elevation models and in computational power have improved the reliability and res-olution of landscape reconstructions significantly. However, no model can be better than the data it relies on and there are uncertainties involved in all the data sources underlying the land uplift model–the lake isolation data, the radiocarbon dating of archaeological data as well as the eustatic model.

There are probably local variations in the land uplift process due to the geological structure of the earth. The data is too sparse to enable reliable detection of this kind of local variation. Also, as far as hydrological features of the landscape are concerned, our current model does not take into account the sedimentation or erosion processes and this might cause inaccuracies when reconstructing the development of riverbeds or lake areas, for example. Despite these shortcomings, landscape development mod-elling remains a valuable tool when interpreting archaeological findings.

References

Dettman D, Mitchell D, Huckleberry G, Foster M (2015) ^{14}C and marine reservoir effect in archae-ological samples from the northeast Gulf of California. Radiocarbon 57(5):785–793. https://doi.org/10.2458/azu_rc.57.18319

Philippsen B (2013) The freshwater reservoir effect in radiocarbon dating. Herit Sci 1(1):1–19. https://doi.org/10.1186/2050-7445-1-24

Reimer R, Reimer P (2006) Marine reservoir corrections and the calibration curve. Pages News 14(3):12–13. https://doi.org/10.1038/nature05214.time

Tiitinen T (2011) Liikettä ajassa ja paikassa—Lounais-Suomen muinaisrannat tarkastelussa. In: Uotila Kari (ed) Avauksia Ala-Satakunnan esihistoriaan. Eura Print Oy, Eura, pp 47–80